Enzymatic Peptide Synthesis

Author
Willi Kullmann, Ph.D.
Senior Research Associate
Institut für Zellbiochemie und Klinische Neurobiologie
Universität Hamburg
Hamburg, West Germany

CRC Press, Inc.
Boca Raton, Florida

Library of Congress Cataloging-in-Publication Data

Kullmann, Willi.
  Enzymatic peptide synthesis.

  Includes bibliographies and index.
  1. Proteolytic enzymes.   2. Proteins--Synthesis.
I. Title.   [DNLM: 1. Peptide Hydrolases.   2. Proteins--
chemical synthesis.   QU 136 K96e]
QP609.P78K84   1987   547.7′5   86-29898
ISBN 0-8493-6841-3

  Direct all inquiries to CRC Press, Inc., 2000 Corporate Blvd., N.W., Boca Raton, Florida, 33431.

© 1987 by CRC Press, Inc.

International Standard Book Number 0-8493-6841-3

Library of Congress Card Number 83-29898
Printed in the United States

PREFACE

Nature has provided an abundant collection of enzymes* which are characterized by their comprehensive, functional diversity. In addition to their "natural" roles this multitude of biocatalysts constitutes an enormous and as yet untapped treasure trove for synthetic organic chemists. Although enzymes had been technically exploited long before their catalytic and chemical essence was fully understood, their use as novel catalysts in organic fine syntheses has been the subject of substantial research only during the last 10 to 15 years. Since there exists an enzyme-catalyzed equivalent for most organic syntheses,[3] one can expect that in the near future a steadily growing number of synthetic applications will emerge from this challenging area of enzyme technology.

One may ask, why do enzymes enjoy a wide and still growing popularity as catalysts for organic syntheses? Certainly, enzymes achieve impressive rate enhancements for the chemical processes they promote. However, the most compelling attraction is based on the unique specificity that they display. From the organic chemist's point of view, therefore, the most notable advantage of enzymes is their capacity to combine, in a concerted fashion, different specificities such as stereo- and regiospecificity thereby opening up new possibilities for degrees of control presently unattainable in any other way. Consequently, the efficient production of chiral synthons with the targeted asymmetric centers and functional groups may be best achieved enzymatically. Another point in favor of enzyme-catalyzed syntheses is that the chemical transformation so mediated takes place under relatively mild conditions, i.e., in a largely aqueous environment and in the absence of potentially toxic solvents, at moderate pH values and temperatures normally ranging from 3 to 10 and 20 to 40°C, respectively, under standard pressure. In summary, the rate accelerations, the incorporation of geometric control into normally random chemical processes, and the mild reaction conditions brought about by enzyme-catalysis significantly aid the preparation of products with a high degree of purity. Simultaneously, the safety factor is increased both for the experimenter and for the environment as a whole.

When searching for enzymes with useful catalytic capabilities for peptide synthetic chemistry, the ability of enzymes to function in vitro in the same fashion as they do in vivo comes into question. Unfortunately, the idea of using enzymes which normally mediate in vivo peptide bond formation in preparative scale peptide synthesis is barely feasible. Apart from other drawbacks they are not commercially available and they are exceedingly difficult to prepare. However, an alternative exists in the proteases which for decades were actually considered to be the true in vivo catalysts of protein biosynthesis.[4] The proteases largely fulfill an essential requirement for routine use in peptide synthetic chemistry in that, for the most part, they can be easily isolated from their natural sources or purchased at reasonable prices. In vivo, proteases serve a variety of different functions.[5] Not only are they the catalysts of generalized protein digestion, they also regulate many biological processes. The feature underlying these physiological activities is their proteolytic capacity which enables the proteases either to decompose a given protein altogether or — by selectively cleaving specific bonds — to stimulate the release of biologically active peptides and proteins from their inactive precursors. To date it has been principally the "destructive" property of the proteases which has been exploited in enyzme technology, in the main, for industrial applicaitons. Thus, proteases were used as early as 1907 — in the leather industry — and over the years a large number of additional uses have been developed. In the meantime, the production of proteases has reached some 500 tons/year,[6] i.e., in terms of quantity, proteases are ahead of all other biocatalysts used in enzyme technology.

---

* The terms enzyme from the Greek ἐν ζύμη (en zyme) for "in yeast" and catalyst from the Greek κατά-λύσις (kata-lysis) for decomposition were coined, respectively, by W. Kühne in 1877 and by J. J. Berzelius in 1835.[1] Moreover, Greek and Arabian alchemists had already gone in search for a catalyst-like substance which they called respectively ξηρῖον (xērion, Engl. desiccative powder) and الكسير (al-iksir, Engl. miraculous mixture).[2] This elusive substance — the much-heralded "philosopher" stone' — was thought to mediate the transformation of base metals to precious metals without itself undergoing change.

Given that under physiological conditons, the equilibrium position in a protease-catalyzed reaction largely favors proteolysis, i.e., the cleavage of peptide bonds, it is not surprising that the proteosynthetic potential of the proteases has been largely neglected. However, according to the principle of microscopic reversibility,[7] the proteases do indeed possess the ability to catalyze the synthesis as well as the hydrolysis of a peptide bond. Consequently, it is the equilbrium point of the reaction, not the nature of the enzyme, that decides whether bonds will be made or broken. Thus, if the equilibrium position of a reaction can be shifted away from proteolysis, then the synthesis of peptide bonds, which is negligible under physiological conditions, may proceed to a significant extent. This is by no means a merely theoretical concept as has been demonstrated by a number of successful protease-mediated peptide syntheses (for reviews see References 8 to 12). Indeed with their special properties of stereo- and regiospecificity and the consequent lack of by-products, the use of protease-controlled reactions may well gain preference over the equivalent chemical syntheses in future artificial peptide and protein preparations.

REFERENCES

1. **Jahn, I., Löther, R., and Senglaub, K.,** Die Entwicklung der Biochemie als einer interdisziplinären Forschungsrichtung der Physiologie und Organischen Chemie, *Geschiechte der Biologie,* Part IV, VEB Gustav Fisher Verlag, Jena, 1982, 498.
2. **Cusumano, J. A.,** Designer catalysts, *Science 85,* 6(a), 120, 1985.
3. **Jones, J. B.,** An illustrative example of a synthetically useful enzyme: horse liver alcohol dehydrogenase, in *Enzymes in Organic Synthesis,* Porter, R. and Clark, S., Eds., Ciba Foundation Symp. 111, Pitman, London, 1985, 3.
4. **Edsall, J. T.,** Some notes and queries on the development of bioenergetics. Notes on some "Founding Fathers" of physical chemistry, J. Willard Gibbs, Wilhelm Ostwald, Walther Nerst, and Gilbert Newton Lewis, *Mol. Cell. Biochem.,* 5, 103, 1974.
5. **Neurath, H.,** Evolution of proteolytic enzymes, *Science,* 224, 350, 1984.
6. **Kula, M.-R.,** Enzyme in der Technik, *Chem. Unserer Zeit,* 14, 61, 1980.
7. **Tolman, R.C.,** *Principles of Statistical Mechanisms,* Oxford University Press, Oxford, 1938, 163.
8. **Jakubke, H.-D. and Kuhl, P.,** Proteasen als Biokatalysatoren für die Peptidsynthese, *Pharmazie,* 89, 1982.
9. **Fruton, J. S.,** Proteinase-catalyzed synthesis of peptide bonds, *Adv. Enzymol. Relat. Areas Mol. Biol.* 53, 239, 1982.
10. **Chaiken, I. M., Komoriya, A., Ohno, M., and Widmer, F.,** Use of enzymes in peptide synthesis, *Appl. Biochem. Biotechnol.,* 7, 385, 1982.
11. **Jakubke, H.-D., Kuhl, P., and Könnecke, A.,** Grundprinzipien der proteasekatalysierten Knüpfung der Peptidbindung, *Angew. Chem.,* 97, 79, 1985.
12. **Kullman, W.,** Proteases as catalysts in peptide synthetic chemistry. Shifting the extent of peptide bond synthesis from a "quantité néglibeable" to a "quantité considérable", *J. Protein Chem.,* 4, 1, 1985.

ACKNOWLEDGMENT

I wish to express my gratitude to Prof. Dr. Bernd Gutte, Department of Biochemistry, University of Zürich, Switzerland, for many stimulating and encouraging discussions. Furthermore, I am indebted to Miss Marion Däumigen for her valuable technical assistance and imaginative design of the illustrations and to Dr. Steve Morley, for critical reading of the manuscript. Finally, I would like to acknowledge the financial support provided by the Deutsche Forschungsgemeinschaft and by the Stifutung Volkswagenwerk.

Willi Kullmann
Hamburg, March 1986

THE AUTHOR

Willi Kullmann, Ph.D., graduated from the University of Cologne in 1975 with a Diplom degree and received the degree Dr. rer. nat. in Biology in 1977 from the same institution. His doctoral thesis dealt with the chemical synthesis of insulin both via classical methods and the Merrifield solid-phase methods.

In 1978, Dr. Kullmann joined the Max Planck Institute for Biophysical Chemistry in Göttingen (FRG) where his basic interest concerned enzymatic peptide synthesis. In the years to follow, he succeeded in preparing a series of neuropeptides by using the enzymatic approach to peptide synthetic chemistry. Currently, Dr. Kullmann is a Senior Research Associate at the Institute for Cell Biochemistry and Neurobiological Clinic at the University of Hamburg (FRG).

TABLE OF CONTENTS

Chapter 14

Chapter 1

INTRODUCTION

The beginnings of peptide synthetic chemistry* can be traced to 1901 when E. Fischer and E. Fourneau reported on the first systematic synthesis of a dipeptide,[1] and further, back to 1882 when T. Curtius unintentionally succeeded in the first in vitro formation of a peptide bond.[2] However, the recent advances in peptide synthetic methodology, cumulating in the fully automatic production of polypeptides, commenced only 30 years ago. The starting point of modern peptide synthetic chemistry was marked in 1953 by the chemical synthesis of the nonapeptide hormone oxytocin by du Vigneaud and his collaborators.[3] Subsequently, further rapid progress has been stimulated principally by the discovery of an ever-increasing number of biologically active peptides. Whereas a review article published in 1953 and entitled *Naturally Occurring Peptides*[4] presented the confirmed structures of only six peptides, the compilation of all currently known peptide structures would grow into a herculean task.

The functional versatility of the peptides is astonishing. Their diverse nature encompasses sweeteners and toxins, antibiotics and ionophores, and chemotactic as well as growth factors. Peptides can act both as stimulators and inhibitors of hormone release. They are involved as morphinomimetics in the pain pathways, and in serving as neurotransmitters, they mediate synaptic communication. They consitute enzyme inhibitors, and conversely, they can function as hormonal messengers to activate appropriate target systems.

The isolation and characterization of a hitherto unrecognized peptide almost invariably entails renewed peptide synthetic activities that may aim at a variety of goals. The classical objective of peptide synthesis is the verification or falsification of primary structures elucidated by sequence analyses of bioactive peptides. Furthermore, synthetic peptides, structurally related to their native counterparts, are effective tools to probe the relationship between peptide structure and biological activity. From a pharmacological point of view it is of particular interest that synthetic analogs of peptide hormones may exhibit unique properties such as superpotency, altered biological specificity, and long-lasting activity. Thus, synthetic peptides may be required for therapeutic purposes, particularly if their natural pendants are not easily obtainable in sufficient quantities.

A novel application of peptide synthetic chemistry is in the production of peptides to be used as immunogens in the generation of antisera specific for proteins of which the peptide represents only a part. Although still in its infancy, this rapidly developing technique not only offers exciting possibilities for raising antigenic determinant specific antibodies but it should also deepen our understanding of the basis of antigenicity. Beyond this lies the creation of entirely artificial peptides with unprecedented primary structures. These truly *ex arte* peptides may be devised either to exhibit putatively nonbiological properties, or to mimic or even enhance commonly known biological activities of their *ex natura* counterparts. Last but not least, peptide synthetic studies may be performed "simply" for the sake of methodological progress.

At present the most frequently used methods of peptide synthesis are those of a chemical nature. The chemical formation of a peptide bond can, in principle, be reduced to four steps (for more detailed descriptions see References 5 to 10; Figure 1).

---

* The term "peptide" as coined by E. Fischer generally denotes unbranched chain-like molecules consisting of up to 100 amino acid residues, whereas molecules possessing more than 100 residues are commonly referred to as proteins. Peptides may be further classified into oligo- and polypeptides; the former containing between 2 and 10 and the latter 11 to 100 amino acid units.

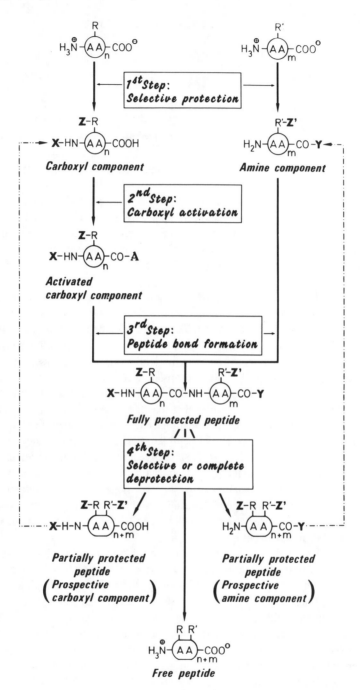

FIGURE 1.   Basic scheme of chemical peptide synthesis. AA, amino acid;
R,R', side-chain functionalities; X, α-amine protecting groups; Y, α-carboxyl
protecting groups; Z,Z', side-chain protecting groups; A, activating substituent.

1.   The main-chain and side-chain functionalities of the educts — amino acids or peptides
     — which are not to participate in the reaction must be selectively protected.
2.   The carboxyl component must be activated. (The carboxyl component and the amine
     component contribute the carbonyl group and the imino group, respectively, to the
     prospective peptide bond.)

3. The peptide bond is formed by coupling the carboxyl component and the amine component via an α-amide linkage.
4. The protecting groups must be removed *in toto*, if the synthesis is completed, or by selective cleavage of the α-amine or α-carboxyl protection, if the synthesis is to be continued.

Originally all synthetic peptides were routinely prepared by conventional solution methods, i.e., the reactants were freely soluble in the reaction media. However, due to B. Merrifield's ingenious innovation of covalently binding the growing peptide chain to an insoluble polymeric support[11] the solid-phase methodology has become increasingly popular over the past 20 years. The most attractive improvement brought about by this procedural simplification is the time-saving and convenient mode of operation and consequently, the prospect of fully automated synthesis.

Numerous peptides have been routinely prepared by solution and solid-phase methods as well as by other techniques such as liquid-phase[12] or alternating solid-liquid phase[13] procedures and remarkable advances have been made in the art of chemical peptide synthesis. Nevertheless, the considerable shortcomings of these methods still impose an "undiminished challenge"[14] upon peptide synthetic chemistry. These limitations arise mainly from the fact that the individual steps of the synthetic pathway are relatively unspecific in nature. Consequently the success of many a synthesis is jeopardized by the appearance of undesired by-products.

To circumvent these problems, an increasing number of organic syntheses is carried out in the presence of enzymes. Due to their stringent specificity, these biocatalysts do not normally permit significant levels of side reactions.

Peptidyltransferase, the enzyme which is responsible for peptide bond formation in vivo, might seem the obvious choice to mediate the enzymatic synthesis of peptides. However, this enzyme is an integral part of the ribosome[15] and its activity is dependent upon the presence of additional ribosomal proteins.[16] Consequently, it is difficult — if not impossible — to isolate this enzyme in a biologically active form.

In contrast, the proteases, which had previously been considered to be the catalysts of protein biosynthesis,[17] are more easily obtainable from their natural sources. Superficially, it may appear incongruous to attempt the synthesis of peptides using proteolytic enzymes. The equilibrium position in a protease-catalyzed reaction is usually far over in the direction of hydrolysis and as a consequence, the reversal of this reaction, i.e., the formation of peptide bonds, represents merely a negligible quantity under physiological conditions. However, the proteases per se cannot be held responsible for this state of affairs. Like other enzymes, they simply accelerate the attainment of equilibrium in a chemical process, whereas the equilibrium point itself, and thus the net synthesis or hydrolysis of the peptide bond, is determined exclusively by thermodynamic factors. Indeed, the proteases do possess the inherent capacity to catalyze both the synthesis and the hydrolysis of a peptide bond, and it is therefore the equilibrium position that actually decides upon the making or breaking of peptide bonds. Consequently, if suitable expedients could be found to shift the equilibrium point of a protease-controlled reaction in favor of the "reverse reaction", the synthesis of peptide bonds may likely become a considerable quantity.

REFERENCES

1. **Fischer, E. and Fourneau, E.,** Über einige Derivate des Glykocolls, *Ber. Dtsch. Chem. Ges.,* 34, 2868, 1901.
2. **Curtius, T.,** Über die Einwirkung des Chlorobenzoyl auf Glycocollsilber, *J. Prakt. Chem.,* 24, 239, 1882.
3. **du Vigneaud, V., Ressler, C., Swan, J. M., Roberts, C. W., Katsoyannis, P. G., and Gordon, S.,** The synthesis of a nonapeptide amide with the hormonal activity of oxytocin, *J. Am. Chem. Soc.,* 75, 4879, 1953.
4. **Bricas, E., and Fromageot, C.,** Naturally occurring peptides, *Adv. Prot. Chem.,* 8, 1, 1953.
5. **Wünsch, E.,** Synthese von Peptiden, in *Methoden der Organischen Chemie,* Vol. 15, Parts 1 and 2, G. Thieme Verlag, Stuttgart, 1974.
6. **Lübke, K., Schröder, E., and Kloss, G.,** Chemie und Biochemie der Aminosären, *Peptide und Proteine,* Vols. 1 and 2, G. Thieme Verlag, Stuttgart, 1975.
7. **Gross, E. and Meienhofer, J.,** *The Peptides: Analysis, Synthesis, Biology,* Vols. 1-3, Academic Press, New York, 1979-1981.
8. **Jakubke, H.-D. and Jeschkeit, H.,** *Aminosäuren, Peptide, Proteine,* Verlag Chemie, Weinheim, 1982.
9. **Bodanszky, M.,** Principles of peptide synthesis, in *Reactivity and Structure: Concepts in Organic Chemistry,* Vol. 16, Hafner, K., Rees, C. W., Trost, B. M., Lehn, J.-M., v.Ragué-Schleyer, P., and Zahradnik, R., Eds., Springer-Verlag, Berlin, 1984.
10. **Bodanszky, M. and Bodanszky, A.,** The practice of peptide synthesis, in *Reactivity and Structure: Concepts in Organic Chemistry,* Vol. 21, Hafner, K., Rees, C. W., Trost, B. M., Lehn, J.-M., v.Ragué-Schleyer, P., and Zahradnik, R., Eds., Springer-Verlag, Berlin, 1984.
11. **Merrifield, R. B.,** Solid phase peptide synthesis. I. The synthesis of a tetrapeptide, *J. Am. Chem. Soc.,* 85, 2149, 1963.
12. **Mutter, M., Hagenmaier, H., and Bayer, E.,** Eine neue Methode zur Synthese von Polypeptiden, *Angew. Chem.,* 83, 883, 1971.
13. **Frank, H. and Hagenmaier, H.,** Ein neues Verfahren zur Peptidsynthese: die alternierende Fest-Flüssig-phasen Methode, *Experientia,* 31, 131, 1975.
14. **Bodanszky, M.,** Peptide synthesis: an undiminished challenge, in *Peptides, Proc. 5th Am. Peptide Symp.,* Goodman, M. and Meienhofer, J., Eds., John Wiley and Sons, New York, 1977, 1.
15. **Weissbach, H. and Pestka, S.,** *Molecular Mechanisms of Protein Biosynthesis,* Academic Press, New York, 1977.
16. **Hampl, H., Schulze, H., and Nierhaus, K. H.,** Ribosomal components from escherichia coli 50S subunits involved in the reconstitution of peptidyltransferase activity, *J. Biol. Chem.,* 256, 2284, 1981.
17. **Florkin, M. and Stotz, E. H., Eds.,** *Comprehensive Biochemistry,* Vol. 32, Elsevier, Amsterdam, 1977, 307.

Chapter 2

REVERSIBLE ZYMO-HYDROLYSIS
A CHRONOLOGY OF ENZYMATIC PEPTIDE SYNTHESIS

The concept of peptide synthesis by reversal of mass action in protease-catalyzed reactions dates back to 1898 when J. H. van't Hoff suggested that the protease "trypsin"* might posses the inherent capacity to catalyze the synthesis of proteins from degradation products originally generated by its own proteolytic action.[2] The rationale behind this idea followed from the applicability of the law of mass action to enzyme-controlled reactions; their reversibility ensuing from the presumed catalytic nature of the process.

The possibility of the participation of hydrolases in not only the degradation but also the assembly of biological macromolecules was first suggested by the phenomenon of the so-called "reversible zymo-hydrolysis", a term introduced in 1898 by A. C. Hill to designate the maltase-catalyzed synthesis of maltose from glucose.[3] Indeed, the earliest reports on glycosidase-, lipase-, and protease-mediated syntheses of glycosides, fats, and peptides were published as early as 1899[3] and 1901,[4,5] respectively, and a further series of studies dealing with enzymatic syntheses of these biomolecules were performed during the following decades (for reviews see References 6 and 7). In contrast, as our present picture of nucleic acid chemistry has evolved only since the early 1950s, it comes as no surprise that the first report of a nuclease-catalyzed formation of oligonucleotides did not appear until 1955.[8]

As mentioned by J. T. Edsall[9] many biochemists, perhaps under the influence of W. Ostwald's "imperative of energetics" (do not waste any energy, but do exploit it),[10] believed that a biochemical process which required free energy to take place could be accomplished with the greatest efficiency by living organisms. As a consequence it was generally taken as granted that the anabolic pathways leading to the synthesis of biological macromolecules were merely reversals of catabolic pathways. This view implicitly suggested that proteins could be prepared by proteolysis in reverse; that is to say, via protease-catalyzed proteo-synthesis. The idea of protein biosynthesis by "reversible enzymic hydrolysis" had for some time been considered to be confirmed by the phenomenon of the so-called "plastein-reaction". The term "plastein" was coined in 1901 by Savjalov to designate the precipitate resulting from the addition of rennin to a partial-hydrolysate of fibrin (peptone).[5] Savjalov, while reproducing an experiment performed by his teacher Danilewski in 1886,[11] correctly identified the plastein formation as the outcome of a "proteo-synthetic" process, namely the reverse of the already known "proteo-lytic" action of proteases. Numerous studies on the subject of plasteins were published in subsequent years, particularly during the first decades of the present century (for further details consult References 6 and 7).

For example, plastein formation was observed upon addition of pepsin, papain, trypsin, and chymotrypsin to concentrated solutions of "peptic" partial-hydrolysates. Although the indications were that the plastein represented a complex mixture of small peptides, their chemical nature remained obscure because the complexity of the digestion mixtures prevented detailed characterization of the plasteins with the methods available at that time. It was not until the 1960s that the mechanism of plastein formation was elucidated by using unambi-

---

* The proteases pepsin and trypsin had already been isolated, respectively, by T. Schwann in 1836 and by W. Kühne in 1877.[1] The designations "pepsin" and "trypsin" were derived from the Greek πεψις (pepsis) for digestion and from τρύχω (trychō) for to wear out, i.e., to digest. In contrast, the denomination "peptide" has its roots in the term "peptone" (a mixture of small peptides generated from proteins via pepsin-catalyzed digestion) the first four letters of which were combined with the last three letters of "polysaccharide", a term derived from the carbohydrate nomenclature.

gously characterized plastein-forming oligopeptides as protease-specific substrates (*vide infra,* Chapter 9).[12]

In 1938, Bergmann and his collaborators were the first to describe the enzymatic synthesis of well-defined peptides. Thus Bergmann and Fraenkel-Conrat succeeded in preparing via papain-catalysis the dipeptide Bz-Leu-Leu-NHPh from benzoylleucine and leucine anilide,[13] and Bergmann and Fruton synthesized Bz-Tyr-Gly-NHPh from benzoyltyrosine and glycine anilide in the presence of α-chymotrypsin.[14] Obviously these studies were inspired by the assumption that in living organisms the protein biosyntheses would be governed by proteases[15] and previously mentioned plastein reactions having provided a stimulus to hypotheses of that kind. However, as the chemical nature of the plasteins had not been exactly described at that time; it was suggested by Bergmann, that the experimental conditions should be basically simplified in order to accomplish unequivocally definable results. For this reason the authors employed — as it is common practice in peptide synthetic chemistry — partially protected starting compounds to enable an exact product analysis. (For these experiments it did not matter that the derivatized products obtained by the enzymatic approach could not be transformed to the free dipeptides.)

The following quotations emphasize how much the synthetic work of Bergmann and in particular of Fruton (for a review see Reference 16), who played a prominent role in these studies in Bergmann's laboratory, was influenced by the idea that the proteases were the protagonists of the in vivo protein synthesis. Bergmann assumed that "... the proteinases owe their existence to the preexistence of other proteinases. There is in life, a practically endless sequence of reactions, in which one proteinase synthesizes the next by a predetermined reaction, and so forth"[15] further to quote Fruton: "The fact that proteolytic enzymes exhibit all these properties in vitro makes it more likely that they play a central role in the course of in vivo synthesis of proteins."[17]

However, the concept of protein biosynthesis by a simple reversal of enzymatic proteolysis, generally accepted as valid until the end of the 1930s, was questioned by thermodynamic data on peptide bond hydrolysis provided by Borsook and Huffman.[18] These authors showed that, under physiological conditions, the synthesis of peptide bonds represents a strongly endergonic process. In fact the formation of a peptide bond requires an energy input of 2 to 4 kcal/mol, and in 1953 Borsook[19] concluded: "Peptide bonds cannot be synthesized to any significant extent merely by mass action reversal of hydrolysis." Furthermore, when in 1941 Lipmann[20] and Kalckar[21] pointed to the salient role of phosphorus compounds such as ATP as energy sources in biochemical processes, it was suggested that the formation of peptide bonds in vivo might proceed by means of "activated" amino acids. Kalckar for instance found: "There is reason to believe that peptide formation in tissues is always coupled with oxidoreduction just like phosphorylation."[21]

Of the various hypotheses promoted during this period to explain protein biosynthesis, the so-called template theory was most popular. According to this view, the activated amino acids align themselves along specific sites of a polynucleotide template and react together resulting in the assembly of a protein. However, the concept of protease-controlled protein biosynthesis was still not dead. As late as 1955, Fruton stated: "It is possible, that our speculations about protein formation are too simple, whether we assume a polynucleotide template or the coupled synthetic action of proteinases."[22]

The demise of the concept of protein biosynthesis via reversible, protease-mediated hydrolysis was finally brought about by the recognition of the genetic code and the crucial roles played by m- and t-RNAs during the process of in vivo protein synthesis (for more details see, for instance, References 23 and 24). After elucidation of the mechanisms of the ribosomal protein biosynthesis, interest in enzymatic peptide synthesis largely waned. However, it spectacularly revived during the second half of the 1970s with the prospect of the utilization of proteases for preparative scale peptide synthesis (*vide infra*).

REFERENCES

1. **Florkin, M. and Stotz, E. H., Eds.,** *Comprehensive Biochemistry,* Vol. 30, Elsevier, Amsterdam, 1972, 265.
2. **Van't Hoff, J. H.,** Über die zunehmende Bedeutung der anorganischen Chemie, *Z. Anorg. Chem.,* 18, 1, 1898.
3. **Hill, A. C.,** Reversible zymohydrolysis, *J. Chem. Soc.,* 73, 634, 1898.
4. **Hanriot, M.,** Sur la réversibilité des actions diastasiques, *C.R. Soc. Biol.,* 53, 70, 1901.
5. **Savjalov, W. W.,** Zur Theorie der Eiweissverdauung, *Pflügers Arch. Ges. Physiol.,* 85, 171, 1901.
6. **Ammon, R.,** Die synthetisierende Wirkung von Fermenten, *Angew. Chem.,* 45, 357, 1932.
7. **Florkin, M., and Stotz, E. H., Eds.,** *Comprehensive Biochemistry,* Vol. 32, Elsevier, Amsterdam, 1977, 307.
8. **Heppel, L. A., Whitfield, P. R., and Markham, R.,** Nucleotide exchange reactions catalyzed by ribonuclease and spleen phosphodiesterase, *Biochem. J.,* 60, 8, 1955.
9. **Edsall, J. T.,** Some notes and queries on the development of bioenergetics. Notes on some 'Founding Fathers' of physical chemistry, J. Willard Gibbs, Wilhelm Ostwald, Walther Nernst, Gilbert Newton Lewis, *Mol. Cell. Biochem.,* 5, 103, 1974.
10. **Ostwald, W.,** Der energetische Imperativ, *Akadem. Verlagsgesellschaft,* 1912.
11. **Danilevski, B.,** The organoplastic forces of the organism (in Russian), in *Comprehensive Biochemistry,* Vol. 32, Elsevier, Amsterdam, 1977, 314.
12. **Wieland, T., Determann, H., and Albrecht, E.,** Untersuchungen über die Plastein-Reaktion. Isolierung einheitlicher Plastein-Bausteine., *Justus Liebigs Ann. Chem.,* 633, 185, 1960.
13. **Bergmann, M. and Fraenkel-Conrat, H.,** The enzymatic synthesis of peptide bonds, *J. Biol. Chem.,* 124, 1, 1938.
14. **Bergmann, M. and Fruton, J. S.,** Some synthetic and hydrolytic experiments with chymotrypsin., *J. Biol. Chem.,* 124, 321, 1938.
15. **Bergmann, M.,** The structure of proteins in relation to biological problems, *Chem. Rev.,* 22, 423, 1938.
16. **Fruton, J. S.,** Proteinase-catalyzed synthesis of bonds, *Adv. Enzymol. Relat. Areas Mol. Biol.,* 53, 239, 1982.
17. **Fruton, J. S.,** Proteases as agents in the formation and breakdown of proteins, *Cold Spring Harbor Symp. Quant. Biol.,* 9, 211, 1941.
18. **Borsook, H. and Huffman, H. M.,** *Chemistry of the Amino Acids and Proteins,* Schmidt, C. L. A., Ed., Charles C Thomas, Springfield, Ill., 1938.
19. **Borsook, H.,** Peptide bond formation, *Adv. Prot. Chem.,* 8, 127, 1953.
20. **Lipmann, F.,** Metabolic generation and utilization of phosphate bond energy, *Adv. Enzymol. Relat. Subj.,* 1, 99, 1941.
21. **Kalckar, H. M.,** The nature of energetic coupling in biological syntheses, *Chem. Rev.,* 28, 71, 1941.
22. **Fruton, J. S.,** Enzymic hydrolysis and synthesis of peptide bonds, *Harvey Lectures 1955,* 51, 64, 1957.
23. **Mahler, H. R., and Cordes, E. H.,** *Biological Chemistry,* 2nd ed., Harper and Row, New York, 1971.
24. **Watson, J. D.,** *Molecular Biology of the Gene,* 3rd ed., Benjamin Inc., New York, 1977.

Chapter 3

THERMODYNAMIC ASPECTS OF PEPTIDE BOND SYNTHESIS

As mentioned in the previous chapter, it is well established that, under physiological conditions, the equilibrium position in protease-catalyzed reactions is far over in the direction of proteolysis. That is to say, the equilibrium constants for the synthesis of free dipeptides are in the range of 10^{-3} to 10^{-4}/mol.[1] Considering these unfavorable auspices, peptide bond formation by mass action via protease catalysis indeed appears to be negligible. Notwithstanding this, consistent with the views of Linderstrøm-Lang[2] and Carpenter,[3] there would be a considerable tendency toward formation of a free dipeptide from its constituent amino acids provided that both educts and product are present in a nonionized form. This idea, however, is easier conceived than achieved. In practice, due to their amphoteric nature, amino acids and peptides are ionized over the entire pH range. They can behave either as cations in a strongly acidic environment or as anions in a strongly basic environment. Furthermore, at pH values between about 3 and 9, they exist in a dipolar, or zwitterionic form, and in the physiological pH region they are almost completely ionized. Under these circumstances, proteolysis will be largely favored over proteosynthesis. Thus the predominant contribution to the energetic barrier to proteosynthesis is accounted for by ionization-neutralization effects. The implication, from the point of view of peptide bond formation at physiological pH values, is that significant thermodynamic work is required for the transfer of a proton from the protonated α-amino group of one reactant to the deprotonated α-carboxyl group of the other reactant.

In a purely formal sense, one might interpret the endergonic process of peptide bond formation from ionized amino acid residues as the sum of the energy-releasing step of peptide bond synthesis from nonionized amino acids and the energy-consuming step of the proton transfer.[4] This may be illustrated by striking a balance of energy with regard to the formation of the dipeptide alanylglycine as shown below (the data are taken from Reference 3):

$$\Delta G_{syn,298}(\text{kcal/mol})$$

$H_2N-CH(CH_3)-COOH + H_2N-CH_2-COOH \rightleftharpoons$		
$\quad H_2N-CH(CH_3)-CO-NH-CH_2-COOH + H_2O$	-3.9	(1)
$H_2N-CH(CH_3)-CO-NH-CH_2-COOH \rightleftharpoons$		
$\quad {}^+H_3N-CH(CH_3)-CO-NH-CH_2-COO^-$	-6.6	(2)
${}^+H_3N-CH(CH_3)-COO^- \rightleftharpoons H_2N-CH(CH_3)-COOH$	$+7.4$	(3)
${}^+H_3N-CH_2-COO^- \rightleftharpoons H_2N-CH_2-COOH$	$+7.3$	(4)
${}^+H_3N-CH(CH_3)-COO^- + {}^+H_3N-CH_2-COO^- \rightleftharpoons$		
$\quad {}^+H_3N-CH(CH_3)-CO-NH-CH_3-COO^- + H_2O$	$+4.2$	(5)

(The contribution of the thermodynamic work required for the proton transfer to the overall free energy change involved in the synthesis of the dipeptide is given by the addition of the above ionization reactions 2 to 4).

As suggested by Carpenter,[3] the free energy change of hydrolysis ($\Delta G_{hyd}°'$) at a given pH can be related to a pH-independent component of the free energy change of hydrolysis ($\Delta G_{hyd}°$) and to an ionization-neutralization term (ΔG_{ion}) in the following way:

$$\Delta G^{o'}_{hyd} = \Delta G^{o}_{hyd} + \Delta G_{ion} \tag{6}$$

or, in terms of the free energy change of synthesis

$$\Delta G^{o'}_{syn} = \Delta G^{o}_{syn} + \Delta G_{ion} \tag{7}$$

By relating the free energy changes to the corresponding equilibrium constants according to the general expression

$$\Delta G^{o} = -RT\ln K \tag{8}$$

Equation 8 can be rearranged to give the pH-dependent equilibrium constant (K_{syn}') for the formation of a peptide bond:

$$K'_{syn} = K_{syn}/K_{ion} \tag{9}$$

In the case of a peptide synthetic reaction, the product and the educts of which are, respectively, fully and partially protected, (a standard procedure in peptide synthetic chemistry), the overall process is

$$
\begin{array}{ccc}
 & K_{syn} & \\
RCOOH + NH_2R' & \rightleftharpoons & RCO - NHR' + H_2O \\
\Updownarrow K_1 \quad\quad \Updownarrow K_2 & & \\
RCOO^- + {}^+NH_3R' & &
\end{array}
\tag{10}
$$

where

$$K_1 = \frac{[RCOO^-][H^+]}{[RCOOH]} \tag{11}$$

and

$$K_2 = \frac{[NH_2R'][H^+]}{[{}^+NH_3R']} \tag{12}$$

are the ionic dissociation constants of the educts, whereas R and R′ represent protecting groups.

The pH-independent equilibrium constant (K_{syn}) reads then in terms of the nonionized form of the reactants

$$K_{syn} = \frac{[RCO - NHR'][H_2O]}{[RCOOH][NH_2R']} \tag{13}$$

The pH-dependent overall equilibrium constant (K'_{syn}) which described the concentrations of all ionic and nonionic forms of the reactants at different pH, obeys the following relation

$$K'_{syn} = \frac{[RCO - NHR'][H_2O]}{[RCOOH + RCOO^-][NH_2R' + {}^+NH_3R']} \tag{14}$$

By inserting the relations of Equations 11 and 12 into Equation 14, the pH-dependent

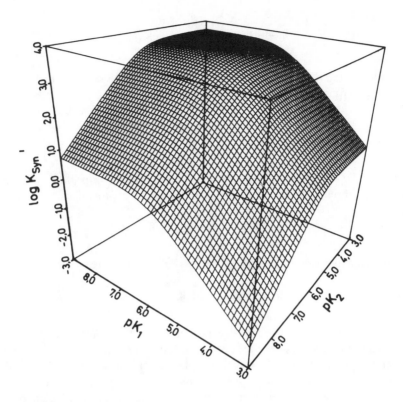

FIGURE 1. Effect of the ionization properties of the educts on the synthesis of a hypothetical peptide of the form RCO-NHR′. The three-dimensional computer representation illustrates the relation between the pH-dependent equilibrium constant K'_{syn} and the dissociation constants K_1 and K_2, as described by Equation 16. The pH value was 6.0 and the logarithm of the pH-independent equilibrium constant K_{syn} was fixed at 3.7, an assumption, which is compatible with the ΔG_{hyd}° values[3] for the hydrolysis of peptides of the form RCO-NHR′.

overall equilibrium constant can be expressed as a function of the pH-independent equilibrium constant:

$$K'_{syn} = \frac{K_{syn}}{\left(1 + \dfrac{K_1}{[H^+]}\right)\left(1 + \dfrac{[H^+]}{K_2}\right)} \tag{15}$$

where the denominator defines the equilibrium constant (K_{ion}) for the above-mentioned proton transfer reaction. (The influence of the ionization properties of the educts upon the peptide bond forming step is illustrated in Figure 1: the three-dimensional presentation is based on the logarithmic form of Equation 15

$$\log K'_{syn} = \log K_{syn} -$$
$$\log(1 + 10^{pH - pK_1} + 10^{pK_2 - pH} + 10^{pK_2 - pK_1}) \tag{16}$$

The optimal pH value for the synthesis can be derived from Equation 15[5]

$$[H^+]_{opt} = \sqrt{K_1 K_2} \tag{17}$$

or
$$pH_{opt} = \frac{1}{2}(pK_1 + pK_2) \tag{18}$$

On substituting the relation of Equation 17 into Equation 15 one obtains the equilibrium constant at the optimal pH value

$$K'_{syn,pH_{opt}} = K_{syn}/\left(1 + \sqrt{\frac{K_1}{K_2}}\right)^2 \tag{19}$$

and when $K_1 \gg K_2$ — an assumption which is usually justified in peptide synthetic studies — then the equilibrium constant at the optimal pH can be approximated[6] by

$$K'_{syn,pH_{opt}} \approx K_{syn}K_2/K_1 \tag{20}$$

The logarithmic expression of Equation 20

$$\log K'_{syn,pH_{opt}} \approx \log K_{syn} - \Delta pK \tag{21}$$

with $\Delta pK = pK_2 - pK_1$, describes the linear dependency of the logarithm of the equilibrium constant of maximal synthesis upon the difference between the dissociation constants. Under the proviso that the reaction in question takes place at or near the optimal pH and that the K_1-value significantly exceeds the K_2-value, ΔpK provides a good measure for evaluating the energy required for the proton transfer, i.e., for estimating the height of the main thermodynamic barrier to peptide bond formation.

REFERENCES

1. **Borsook, H.,** Peptide bond formation, *Adv. Prot. Chem.,* 8, 127, 1953.
2. **Linderstrøm-Lang, K.,** Biological synthesis of proteins, Lane medical lectures 1951, in *Selected Papers,* Academic Press, New York, 1962, 448.
3. **Carpenter, F. H.,** The free energy change in hydrolytic reactions: the non-ionized compound convention, *J. Am. Chem. Soc.,* 82, 1111, 1960.
4. **Waldschmidt-Leitz, E.,** On the enzymatic synthesis of peptide bonds, *Angew. Chem.,* 61, 437, 1949.
5. **Dobry, A., Fruton, J. S., and Sturtevant, J. M.,** Thermodynamics of hydrolysis of peptide bonds, *J. Biol. Chem.,* 195, 149, 1952.
6. **Yagisawa, S.,** Studies on protein semisynthesis. I. Formation of esters, hydrazides, and substituted hydrazides of peptides by the reverse reaction of trypsin, *J. Biochem.,* 89, 491, 1981.

Chapter 4

FORMATION OF PEPTIDE BONDS — PROTEASES AS "ACTIVATING SYSTEMS"

The foregoing discussions on the thermodynamics of protease-catalyzed reactions address the basic problem of where the chemical reactions are going; that is, they dealt with the problem of equilibrium or chemical statics. Although thermodynamic studies may answer this salient question, they cannot predict the rate of attainment of equilibrium. The latter question is a problem of the reaction rate or of chemical kinetics. Even the hydrolysis of a peptide bond, the highly exergonic nature of which under physiological conditions was described in the preceeding chapter, represents a rather sluggish process in the absence of a catalyst. The reaction rate of peptide hydrolysis is barely measurable within a reasonable time interval, thus while peptides, in an aqueous environment, can be considered to be thermodynamically unstable, nevertheless in a kinetic sense they are rather stable.

The rationale behind this reaction inertia is as follows: before a chemical reaction, such as the synthesis or hydrolysis of a peptide bond, can take place, the molecules have to reach a transition state, i.e., a state of higher potential energy relative to the lower potential energies of the initial and final state (Figure 1). Prior to reaction therefore, the molecules have to acquire a critical energy, the so-called "activation energy", in order to attain the energy-rich transition state where chemical bonds are in the process of being formed or broken. (Figure 1 provides a pictorial idea of the activation energy as the potential-energy hill that must be climbed to arrive at the activated state.)

During an uncatalyzed chemical reaction, however, only a very limited number of molecules possess sufficient kinetic energy of their own to arrive at the top of the activation barrier; accordingly, the reaction rate is very low.

In a unimolecular process such as

$$X_o \rightarrow (X_A) \rightarrow \text{products}$$

where the transition state, X_A, and the ground state, X_o, are in a thermodynamic equilibrium, the equilibrium constant for the formation of the activated state is given by:

$$K_A = \frac{[X_A]}{[X_o]} \tag{1}$$

and the reaction rate is as follows:

$$\frac{-d[X_o]}{dt} = k_1[X_o] = \nu_A[X_A] \tag{2}$$

Here k_1 represents the rate constant for the decay of X_o and ν_A constitutes the average frequency with which the activated transition state decomposes into the products. According to Eyring,[1] ν_A can be substituted by the universal frequency factor kT/h (where k and h are the Boltzmann and Planck constant, respectively), and the rate constant, k_1, can be defined with regard to Equation 1 as

$$k_1 = \frac{kT}{h} K_A \tag{3}$$

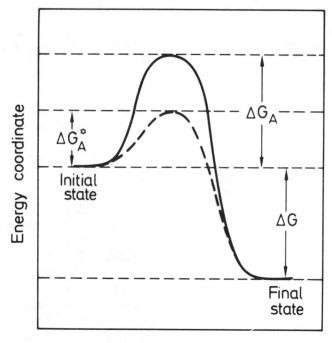

FIGURE 1.   Energy barrier surmounted by a system in an uncatalyzed (——) and a catalyzed (- - -) chemical reaction. ΔG indicates the overall free energy of the reaction, whereas ΔG_A and $\Delta G_A{}^*$ denote the energy of activation in the absence and in the presence, respectively, of a catalyst.

Since (cf. Equation 8, Chapter 3)

$$K_A = \exp(-\Delta G_A/RT) \tag{4}$$

where ΔG_A is the activation energy, Equation 3 becomes

$$k_1 = (kT/h)\exp(-\Delta G_A/RT) \tag{5}$$

By using another relationship from equilibrium thermodynamics,

$$\Delta G_A = \Delta H_A - T\Delta S_A \tag{6}$$

the activation energy, ΔG_A, can be partitioned between an enthalpic and an entropic term and Equation 3 becomes

$$k_1 = (kT/h)\exp(\Delta S_A/R)\exp(-\Delta H_A/RT) \tag{7}$$

where the quantities ΔH_A and ΔS_A are called, respectively, the enthalpy and entropy of activation. (To be more rigorous, the expression for k_1 should be multiplied by a factor k, known as the transmission coefficient, which describes the probability that the transition-state complex decomposes into the products instead of back into the educt. However, this factor is generally close to unity and may be neglected.)

   Hence, the reaction rate is not only dependent upon the temperature and the concentration of the transition state species, being also a function of the activation energy, i.e., the enthalpy

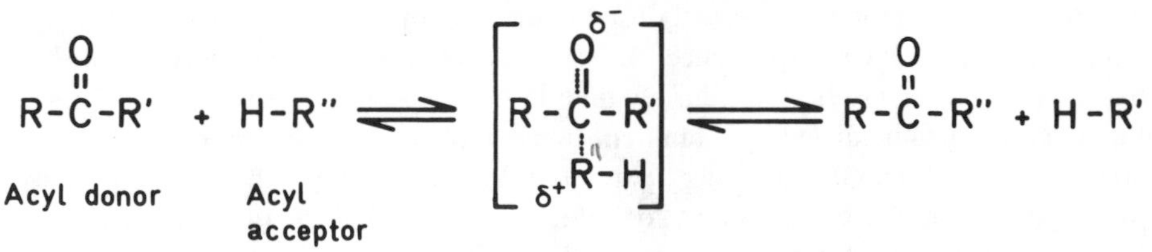

$$\underset{\substack{\text{Acyl donor}}}{\text{R}-\overset{\overset{\textstyle O}{\|}}{\text{C}}-\text{R}'} \; + \; \underset{\substack{\text{Acyl}\\\text{acceptor}}}{\text{H}-\text{R}''} \;\; \rightleftharpoons \;\; \left[\overset{\overset{\textstyle O^{\delta-}}{\|}}{\underset{\underset{\textstyle \delta+\text{R}-\text{H}}{|}}{\text{R}-\text{C}-\text{R}'}} \right] \;\; \rightleftharpoons \;\; \text{R}-\overset{\overset{\textstyle O}{\|}}{\text{C}}-\text{R}'' \; + \; \text{H}-\text{R}'$$

FIGURE 2.   Acyl group transfer.

and entropy of activation. By definition, a catalyst enhances the rate of reaction by lowering the activation energy; a process that allows a greater number of molecules to enter an energy-rich state. In the presence of a biocatalyst, biochemical reactions can be accelerated by greater than 10 orders of magnitude. The efficiency of enzymatic catalysis may be illustrated by the following numerical example: an increase of a reaction rate by a factor of 10^{10} reduces the half-life time of the reaction from originally 300 years to 1 sec.[2]

The enzymes catalyzing the processes in which peptide synthetic chemists are most interested are the proteases. Although commonly known for their proteolytic action, they obey, in common with all other biocatalysts, the principle of microscopic reversibility. As a consequence, proteases accelerate both the forward and the reverse reaction of the protease-specific chemical processes to the same degree. Since an enzyme changes rate but not equilibrium, the position of the latter exclusively determines whether a particular peptide bond is preferentially generated or destroyed in a given protease-catalyzed reaction.

In chemical terms, the formation or cleavage of a peptide can be considered as an electrophilic acylation of — or an acyl group transfer to — an α-amino group of an amino acid or a water molecule, respectively.

The common, uncatalyzed reaction of an acyl group donor in the form of RCO-R' (R' = OH and NH − R, respectively) with an acyl group acceptor such as H − R'' (R'' = NH − R and OH, respectively) leads to a transition state which is characterized by the development of a positive charge on the attacking nucleophilic acceptor and a negative charge on the carbonyl oxygen of the acylating compound (Figure 2). A transition state of that kind is highly unfavorable from an energetical point of view because of the existence of labile positive and negative charges. However, a stabilization of these charges would lower the energy of the transition state. Consequently, the activation energy for the acyl transfer would be reduced and the reaction would proceed more rapidly. The addition of a base, for example, that withdraws a proton from the nucleophile, could stabilize the positive charge, while the negative charge would be stabilized in the presence of an acid, i.e., of a proton donor. Both these procedures, general base, as well as general acid catalysis, are routinely used for the hydrolysis of peptides and proteins. Alternatively, an energetically more favorable transition state can also be obtained if, prior to reaction, either the nucleophilicity of the acyl group acceptor or the electrophilicity of the α-carboxyl carbon of the acyl group donor is increased. The second alternative is generally adopted in peptide synthetic chemistry, that is to say the electron density of the α-carboxyl carbon is decreased by introducing an electronegative substituent R' (cf. Figure 2) which either replaces a hydrogen or a hydroxyl group of the α-carboxyl moiety. This so-called "carboxyl-activation" provides an energy-rich intermediate of high acyl group transfer potential which readily undergoes aminolysis or hydrolysis, thus generating or destroying peptide bonds.

The principle of carboxyl activation is ubiquitous in the field of peptide or protein synthesis both in vivo and in vitro. A typical species of an activated acyl group donor is the so-called active ester of the form RCOO − R' (R' = alkyl, acyl, aryl). In ribosomal protein synthesis the peptidyl-t-RNA represents an active peptide ribosyl ester which exhibits strong acylating power, and, as a consequence, is readily aminolyzed by an incoming amino acyl-t-RNA.[3]

This ester type activation is also widely used in the field of chemical peptide synthesis. (For a comprehensive review see Reference 4). Carboxyl activation is not only accomplished via active oxygen-based esters; thus during nonribosomal peptide biosynthesis of microbial antibiotics, which is mediated by a multienzyme complex, and to a minor extent during the chemosynthesis of peptides, thioesters play the role of an energy-rich acyl-group donor.[4,5]

What strategy do proteases adopt to reduce the activation energies for the reactions catalyzed by them? Furthermore, how can the molecular mechanisms of protease-action account for the remarkably large rate-accelerations observed during enzymatic peptide bond hydrolysis or synthesis?

Answers to these questions may be deduced from studies on the protease chymotrypsin, one of the best-characterized enzymes of all. The molecular mechanisms governing the chymotryptic activities are understood at least at a basic level. Thus, the biologically active form of the protease is post-translationally manufactured by tryptic cleavage of the arginyl$_{15}$-isoleucine$_{16}$ peptide bond of chymotrypsinogen. This enzymatically inactive precursor represents a single-chain protein consisting of 245 amino acid residues. Additional autocatalytic cleavage steps lead to a diverse range of biologically active variants of chymotrypsin, the most widely known of which is α-chymotrypsin — a triple-chain molecule whose subunits are cross-linked via disulfide bridges. The "chemistry" of chymotryptic catalysis not only provides an insight into the mode of action of proteases but it also describes the factors essential for the catalytic efficiency of an enzyme. In the following paragraph, the molecular events and forces operative during the chymotryptic catalysis will be described. (The individual steps of the proposed catalytic pathway[6] are outlined in Figure 3.)

The key amino acids at the active site of chymotrypsin constitute a "charge relay system" comprising the side-chains of aspartic acid$_{102}$, histidine$_{57}$, and serine$_{195}$ which, due to their unique juxtaposition, can be linked by hydrogen bonds. Through a concerted acid-base catalysis mediated by this "triad" a proton can be withdrawn from the hydroxyl group of the serine residue and then be shifted via the imidazole ring-system of the histidine to the β-carboxylate function of the aspartic acid unit. This process is reversible. Thus as a result of the forward reaction the highly reactive alkoxide anion of serine is created, whereas the third functionality of aspartic acid is deprotonated during the reverse reaction, i.e., the initial state is restored. Located in the immediate vicinity of this active site is the substrate-binding site, the molecular architecture of which resembles a deeply invaginated cavity rich in hydrophobic amino acid residues. The narrowness of the binding pocket leaves just enough space to accomodate a benzyl-, phenyl-, or indole group. The binding site therefore appears to be "tailor-made" for the side-chain moieties of the aromatic amino acids such as phenylalanine, tyrosine, or tryptophan. Consequently, its three-dimensional structure determines the primary specificity (*vide infra*, Chapter 6, Section A.1) of chymotrypsin, whereas its spatial arrangement relative to the entire molecule defines the stereospecificity of the protease.

When a given protein, peptide, or amino acid derivative fulfills the criteria characterizing a substrate of chymotrypsin, then its aromatic side-chain becomes anchored in the substrate binding cavity via a multitude of hydrophobic interactions. Beyond that, the mobility of the substrate is further restricted by the spatial constraints in the direct neighborhood of the binding area, by hydrogen bonds which attach the imino group and the carbonyl oxygen of the aromatic amino acid to the peptidic backbone of the enzyme, and by contacts of hydrophobic nature between the protease and the presumptive "leaving group", R', (cf. Figure 3) of the substrate. As a result, the substrate is fixed into a position that enables the enzyme to attack the reactive bond. As such the enzyme and substrate have been assembled to form the so-called Michaelis complex (Figure 3a).

In the next step of the catalytic pathway, the deprotonated oxygen, O^γ, of the serine$_{195}$ side chain rotates around its $C^\alpha - C^\beta$ bond and begins to contact the trigonally coordinated carbonyl carbon, C', of the substrate. Simultaneously, the C'-atom moves out of its plane

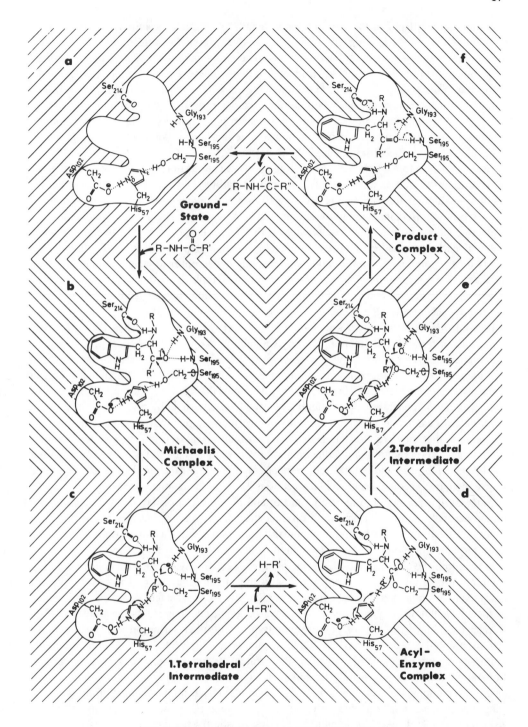

FIGURE 3.   Proposed pathway of chymotryptic catalysis.

of coordination defined by the three original ligands and towards the O^γ-atom. On closer approach to O^γ, a covalent $C' - O^\gamma$ bond will be formed and a tetrahedrally coordinated transition state comes into being which has a centrally positioned C'-atom and a negatively charged carbonyl oxygen that is optimally aligned toward its hydrogen-bond donors (Figure 3b). The relatively unstable tetrahedral conformation has the tendency to collapse very rapidly either by the expulsion of the attacking new ligand, O^γ, so that the initial state is restored, or by the release of the leaving-group, R', i.e., a novel trigonally coordinated state emerges, the so-called acyl-enzyme complex (Figure 3c). The cleavage of the leaving group is initiated by a minor dislocation of the side chain of histidine$_{57}$ thereby enabling the transfer of the N^ϵ-proton of the imidazole system to the leaving group. This process is paralled by a proton transfer from the β-carboxylate of the aspartic acid$_{102}$ residue to the ring nitrogen, N^δ, of His$_{57}$. As a consequence, the imidazole ring system is electrically neutralized and the deprotonated aspartic acid side chain is again available to act as a proton acceptor (Figure 3c).

In this situation, the proton of a given acyl-group donor, $H-R''$, can be attached to the deprotonated N^δ-atom of the imidazole ring — as a counter-move, the proton of the N^ϵ-atom of the imidazole is again transferred to the β-carboxylate anion of the aspartic acid$_{102}$ residue — whereas the "incoming-group", R'', starts a nucleophilic attack toward the C'-atom of the acyl-enzyme complex (Figure 3d). When this attack is successful and the respective atoms are brought close enough so that a covalent bond begins to form between them, a second tetrahedral intermediate develops. Upon cleavage of its C'-O^γ linkage the newly formed tetrahedral conformation subsequently decomposes to yield a product in which the original leaving group of the educt has been finally replaced by the incoming group, R'' (Figure 3e). These molecular events coincide with a proton transfer from Asp$_{102}$ via His$_{57}$ to Ser$_{195}$ which restores the initial state of the active site (Figure 3f).

It would be informative to know in what way chymotrypsin obviates the difficulties associated with the development of the energetically unfavorable tetrahedral transition state which characterizes the nonenzymatic reaction (Figure 2). We have seen that the chemical reaction requires a considerable activation energy which constitutes a thermodynamic barrier to the progress of the reaction. In fact the enzyme avoids the formation of this labile transition state. Instead it generates two differently shaped energy-rich intermediary states which nevertheless bear some structural resemblance to the original transition state outlined in Figure 2, in that they are also tetrahedrically coordinated and develop, at least temporarily, two charges. These novel intermediary states are energetically more favorable because their charges are stabilized and neutralized, respectively, *in statu nascendi*. This is accomplished via chymotrypsin-mediated acid and base catalysis by using the α-imino group of the glycine$_{193}$ residue as a proton donor (Figures 3c and e) and the N^ϵ-atom of the histine$_{57}$ residue as a proton acceptor (Figures 3b and d). Consequently, the activation energies required for the generation of the chymotrypsin-based transition states are significantly reduced relative to those required for the uncatalyzed formation of the original transition state.

In addition, the above-mentioned principle of carboxyl activation is also achieved during chymotrypsin-catalyzed acyl group transfer to the targeted acceptor. Thus in the present case the energy-rich intermediate exhibiting a high acyl group transfer potential is represented by the covalent acyl-enzyme complex (Figure 3d). As the enzyme and its substrate are linked to each other through an ester bond, chymotryptic and other serine protease-mediated reactions proceed via an "active" ester in a manner analogous to many chemical peptide syntheses. However, the dual transition state strategy adopted by chymotrypsin to reach its "operative point" only partially explains the advantages of an enzymatic catalysis. For example a model-system designed by Anderson et al.[7] which included some of the chemical "ingredients" essential for chymotrypsin catalysis, such as N^α-acetylserine amide, imidazole, and N^α, O^γ-diacetylserine amide acting as substitutes for the enzyme-based serine$_{195}$ and histidine$_{57}$ residues as well as for the acyl-enzyme complex, demonstrated significant rate accelerations of both the acylation and the deacylation step relative to the uncatalyzed reactions. However, these features are by no means sufficient to provide for the efficiency of catalysis which characterizes the native enzyme. Thus what are the enzyme-specific properties that establish the superiority of native chymotrypsin over the above model system?

It has been widely interpreted that the uncatalyzed reaction represents an intermolecular process whereas the enzyme mediates an intramolecular reaction which proceeds per se more rapidly because the reactants are bound to the enzyme. Certainly, the intramolecular nature of an enzyme-catalyzed reaction contributes crucially to its efficiency. However, Anderson et al.[7] showed that this cannot fully account for the superiority of enzymatic catalysis. Thus the acyl group transfer potential of a urea-denatured O^γ-acetylserine$_{195}$-chymotrypsin did not exceed that of the aforementioned model-substance, N^α, O^γ-diacetylserine amide, under otherwise identical conditions. Whether in the native or denatured state the enzyme will retain the chemical functions involved in the catalytic mechanism. Evidently, the structural conformation of the enzyme and therefore the mutual orientation of the chemical functions participating in catalysis must play a crucial role in promoting the enhanced catalytic potential of the native enzyme over the denatured enzyme. One might now consider the significance of the correct orientation of these functional groups in relation to the reaction pathway outlined in Figure 3.

The spatial arrangement of the triad Ser$_{195}$-His$_{57}$-Asp$_{102}$ (Figure 3) enables a concerted and, consequently, extremely rapid proton transfer along hydrogen bonds starting from the serine$_{195}$ residue via the histidine$_{57}$ residue, which can act both as proton donor and acceptor due to its imidazole-based nitrogen atoms, to the aspartic acid$_{102}$ residue, and as occasion demands, vice versa. Presumably the anionic form of the β-carboxylate group of the aspartic acid plays a role in inducing a positive charge on the adjacent N^δ-atom of the imidazole ring, whereupon, the partially negative charge which develops on the remote N^ϵ-atom exerts a ''suction effect'' on the proton of the nearby hydroxyl group of the serine$_{195}$ residue, thus further promoting the formation of the highly reactive alkoxide anion. As a matter of course, the extraordinary catalytic performance of the so-called charge-relay system in the active site of chymotrypsin is crucially dependent on the correct order of succession, the optimal alignment, and on the immediate environment of the constituents of the systems, i.e., of the side chains of the individual amino acid residues. Upon denaturation of the enzyme, however, the ingeniously ''tailored'' triad will decay, and the catalytically active groups — from now acting as ''soloists'' — will perform less efficiently.

The same comes true for the substrate-binding site, the architecture of which consists of a number of amino acid residues. Although not juxtaposed within the primary sequence, these amino acids are located in close proximity to each other within the tertiary structure of chymotrypsin. In the presence of a denaturing agent the ordered three-dimensional structure of the protein is unfolded and as a result not only the geometry of the active site but also the substrate-binding region will be destroyed. Consequently, the disordered enzyme loses the capacity, which, in fact, the model-system of Anderson et al.[7] never possessed, to bring the reactants together from a dilute solution and to coordinate them in close vicinity to each other and to the active site in such a way that their reactive groups are optimally aligned for the desired reaction. Due to their ''immediate neighborhood'', which implies a remarkable increase in their ''local'' concentration, the collision frequency between the reactants is significantly greater than in dilute solutions. Futhermore, the fixation and orientation of the reactants ensures systematic rather than the random collisions that could be expected from molecules freely mobile in solution.

The significance of the manner in which a structurally intact chymotrypsin molecule concentrates and orients its substrates is underlined by the following calculus: For the chymotrypsin-catalyzed reaction outlined in Figure 3 to take place, five different moieties must collaborate, namely two reactants and three catalytic groups. In the presence of denatured chymotrypsin, the velocity of a fifth-order reaction would be extremely small, whereas a rate enhancement of 13 orders of magnitudes could be expected,[8] if the five moieties could be assembled in the active site of the enzyme, i.e., if the reaction were de facto made to be first order.

As already mentioned above (cf. Chapter 4, Equation 4), part of the activation energy is enthalpic and another part is entropic in origin. In principle, the chymotrypsin-induced concentration effects and the correct orientation of the substrate may be considered as contributions to a change in the activation entropy, whereas the alteration in charge distribution at the active site and the geometry of the substrate-binding area affect the activation enthalpy of the transition state. In spite of the above, it is difficult or even impossible to draw a distinct dividing line between enthalpic and entropic effects on a molecular level. The aforementioned transition state generated during chymotryptic catalysis (Figure 3) may be used as an illustration; thus upon the formation of the tetrahedral intermediate, strain is imposed on the original bond angles of the trigonal form and the atoms involved are compressed,[6] i.e., processes, that are manifested in enthalpy changes. On the other hand, the mobility of the groups concerned with the transition state is restricted, i.e., a process that is accompanied by entropy changes.

The catalytic pathways followed by other serine proteases such as trypsin,[9,10] subtilisin,[6,11] and probably carboxypeptidase Y[12] appear similar to that described for chymotrypsin. Thus, in analogy to the mode of activation of chymotrypsin, their substrates are also covalently bound to a serine residue to form an O-acyl complex. In contrast, the cysteine protease papain covalently binds its substrates — also by imidazole-aided catalysis (*vide supra*) — through an "activated" thio-ester.[13,14] Papain is involved in the resulting S-acyl complex via the thiol-group of a cysteine residue; a mode of activation that is most probably also favored by other cysteine proteases such as ficin and bromelain.[14]

In addition to the above mechanisms, a highly reactive, covalent acyl-enzyme complex can also be generated via an anhydride linkage between the β- and γ-carboxyl groups, respectively, of a protease-inherent amino dicarboxylic acid residue of a protease and the α-carboxyl group of its respective substrate. This type of activation is analogous to a technique frequently employed during chemical peptide syntheses, the so-called mixed anhydride method. Proteases which possibly favor this mechanism are pepsin,[15-17] carboxypeptidase A[18,19] and, *cum grano salis,* thermolysin.[20]

The activated acyl-protease complexes of both the ester and the anhydride type are at a high energy level and can therefore readily undergo deacylation by a nucleophilic attack. Frequently the nucleophile is a water molecule and the activated complex is hydrolyzed, i.e., the acyl group is transferred to the water molecule, and the commonly known proteolysis takes place (Figures 4a and b). Less commonly, where the nucleophile is the α-amino group of an amino acid residue, the activated complex is aminolyzed, i.e., the acyl group is transferred to the amino acid, and synthesis of a peptide bond takes place (Figures 4a and b). In fact, the partition of the acyl-protease complexes between hydrolytic and aminolytic cleavage does not only depend on the concentrations of the deacylating agents and consequently peptide bond synthesis is not as unfavored as it may first appear. If the opposite were true then the rate of proteolysis would of course significantly outstrip proteosynthesis in an aqueous solution. However, the ratio of hydrolysis to aminolysis is also strongly influenced by the nucleophilic strength of the competitors; and in this respect the amino acids — especially C-terminally protected ones — have an unequivocal advantage over water. To give an example: due to its strongly nucleophilic character, the leucine amide (0.25 M) is able to deacylate an acyl-chymotrypsin complex 20 times faster than water (55 M).[21] As a consequence, in this case proteosynthesis would largely predominate over proteolysis in a kinetically controlled reaction (cf. Chapter 5, Section III).

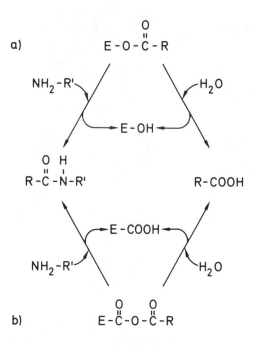

FIGURE 4. Enzymatic carboxyl activation via (a) an "active ester" or (b) a "mixed anhydride", and subsequent deacylation of the resulting acyl-protease complex by aminolysis or hydrolysis, respectively.

REFERENCES

1. **Eyring, H.,** The activated complex and the absolute rate of chemical reactions, *Chem. Rev.,* 17, 65, 1935.
2. **Page, M. J.,** Entropy, binding energy, and enzymic catalysis, *Angew. Chem. Int. Ed. Engl.,* 16, 449, 1977.
3. **Weissbach, H. and Pestka, S.,** *Molecular Mechanisms of Protein Biosynthesis,* Academic Press, New York, 1977.
4. **Wünsch, E.,** *Methoden der Organischen Chemie,* Vol. 15, Part II, G. Thieme Verlag, Stuttgart, 1974.
5. **Lipmann, F.,** Nonribosomal polypeptide synthesis on polyenzyme templates, *Acc. Chem. Res.,* 6, 361, 1973.
6. **Blow, D. M.,** Structure and mechanisms of chymotrypsin, *Acc. Chem. Res.,* 9, 145, 1976.
7. **Anderson, B. W., Cordes, E. H., and Jencks, W. P.,** Reactivity and catalysis in reactions of the serine hydroxyl group and of O-acyl serines, *J. Biol. Chem.,* 236, 455, 1961.
8. **Koshland, D. E., Jr.,** The comparison of non-enzymic and enzymic reaction velocities, *J. Theoret. Biol.,* 2, 75, 1962.
9. **Keil, B.,** Trypsin, in *The Enzymes,* 3rd ed., Part III, Boyer, P. D., Ed., Academic Press, New York, 1971, 249.
10. **Huber, R. and Bode, W.,** Structural basis of the activation and action of trypsin, *Acc. Chem. Res.,* 11, 114, 1978.
11. **Robertus, J. D., Kraut, J., Alden, R. A., and Birktoft, J. J.,** Subtilisin; a stereochemical mechanism involving transition-state stabilization, *Biochemistry,* 11, 4293, 1972.
12. **Auld, D. S.,** Direct observation of transient ES complexes: implications to enzyme mechanisms, in *Bioorganic Chemistry,* Vol. I, van Tamelen, E. E., Ed., Academic Press, New York, 1977, 1.
13. **Bender, K. L. and Brubacher, L. J.,** The kinetics and mechanisms of papain-catalyzed hydrolysis, *J. Am. Chem. Soc.,* 88, 5880, 1966.
14. **Lowe, G.,** The cysteine proteinases, *Tetrahedron,* 32, 291, 1976.

15. **Newmark, A. K. and Knowles, J. R.,** Acyl- and amino-transfer routes in pepsin-catalyzed reactions, *J. Am. Chem. Soc.,* 97, 3557, 1975.
16. **Takahashi, M. and Hofmann, T.,** Acyl intermediates in penicillopepsin-catalyzed reactions and a discussion of the mechanisms of action of pepsins, *Biochem. J.,* 147, 549, 1975.
17. **Wang, T.-T. and Hofmann, T.,** Effects of secondary binding by activator peptides on covalent intermediates of pig pepsin, *Biochem. J.,* 153, 701, 1976.
18. **Kaiser, B. L. and Kaiser, E. T.,** Carboxypeptidase A: a mechanistic analysis, *Acc. Chem. Res.,* 5, 219, 1972.
19. **Lipscomb, W. N.,** Carboxypeptidase A mechanisms, *Proc. Natl. Acad. Sci. U.S.A.,* 77, 3875, 1980.
20. **Morihara, K., Tsuzuki, H., and Oka, T.,** Acyl and amino intermediates in reactions catalyzed by thermolysin, *Biochem. Biophys. Res. Commun.,* 84, 95, 1978.
21. **Morihara, K., and Oka, T.,** α-Chymotrypsin as the catalyst for peptide synthesis, *Biochem. J.,* 163, 531, 1977.

Chapter 5

PROTEOSYNTHESIS VS. PROTEOLYSIS: HOW TO BIAS THIS ANTAGONISM IN FAVOR OF PROTEOSYNTHESIS

ABSTRACT

From thermodynamic considerations (*vide supra*), one can readily deduce that under physiological conditions the equilibrium in protease-catalyzed reactions lies far over in the direction of peptide bond hydrolysis. In order to enforce peptide bond synthesis in spite of the above, one has to resort to some expedients capable of either reducing or bypassing the energetic barriers to the reversal of hydrolysis. In the following several possible strategies are described aiming at an improvement of the *a priori* rather unfavorable prospects of peptide synthesis at the expense of peptide hydrolysis.

I. SHIFT OF IONIC EQUILIBRIA

The main obstacle to peptide bond synthesis, which comes in form of the energy required for the previously mentioned proton transfer, depends crucially on the ionization equilibria of the reactants. From Equation 21 (see Chapter 3), it is at once apparent that an increase of the pK_1 value and/or a decrease of the pK_2 value of a given pair of educts, will reduce the energy consumption of the proton transfer and will, as a result, cause the equilibrium shift in favor of peptide synthesis.

A. Cancelling the Zwitterionic Character of the Reactants

The highly endergonic characters of the proton transfer can be accounted for by the strong acidity and basicity of the α-carboxyl- and α-amino groups of the reactants which in turn, can be attributed to the zwitterionic nature of free amino acids and peptides. Consequently, a favorable pK shift of the ionogenic groups may be achieved by curtailing the zwitterionic character of the respective educts. This can be readily done by the introduction of α-amino- and α-carboxyl-protecting groups, a procedure which is routinely used during chemical peptide syntheses. The consequences of this manipulation are illustrated by the fact that the energy required for the proton transfer during the synthesis of the dipeptide glycylglycine from the constituent free amino acid amounts to 8.2 kcal/mol at pH 7.[1] However, if protected glycine residues such as N^α-benzoylglycine and glycine amide could react with each other to give N^α-benzoylglycylglycine amide, the energy input for the proton transfer would come to only 5.6 kcal/mol at the optimal pH value (in the present case, 5.9).

The free energy change of hydrolysis would still be negative ($\Delta G^{o\prime}_{hyd} = -1.7$ kcal/mol), i.e., the equilibrium position of this reaction is still over in the direction of the cleavage products; relative to the initial state ($\Delta G^{o\prime}_{hyd} = -3.6$ kcal/mol), however, the energy balance is substantially altered in favor of synthesis.

B. Effects of Organic Co-Solvents

Another technique widely applied to shift ionic equilibria in favor of peptide synthesis is to add organic solvents to the reaction mixture. By lowering the dielectric constant of the reaction medium, the organic co-solvent diminishes the hydration of ionic groups. This effect predominantly affects the pK value of those groups whose ionization represents a separation of charge;[2] most notably carboxyl functions. Thus, the acidity of the α-carboxyl group preferentially and, to a lesser extent, the basicity of the α-amino group can be reduced.[3,4] The difference between the acidity of an ionizable group in an aqueous and an aqueous-organic mixed solution (ΔpK_a) can be described by the following empirical relation:[5]

$$\Delta pK_a = \frac{36{,}000 \cdot Z}{T} \left(\frac{1}{D_w} - \frac{1}{D_{w-o}} \right) \tag{1}$$

where D_w and D_{w-o} are the dielectric constants of the aqueous and of the aqueous-organic solutions, respectively, whereas Z represents the valency of the proton accepting form, this corresponding to the base in the Brønsted concept.

C. Variation of Temperature and pH

Ionic equilibria can also be perturbated in favor of peptide synthesis by elevation of the reaction temperature. The ionization behavior of acids and bases as a function of temperature can be generally expressed by the following equation:[6]

$$-\frac{d(pK)}{dT} = \frac{pK}{T} + \frac{\Delta S^o}{2.303RT} = \frac{pK + 0.218\Delta S^o}{T} \tag{2}$$

The degree of ionization of the α-amino groups of the amino acid varies significantly with temperature (the pK_2 value decreases with increasing temperature by roughly 0.025 to 0.030 pH units per °C) whereas that of the α-carboxyl function depends only slightly on temperature.

In fact, the ionization properties are also influenced by the hydrogen ion concentration of the reaction medium. However, any pH variation is a double-edged sword, since any favorable shift of the ionic equilibrium of one educt is accompanied by an unfavorable shift of the ionic equilibrium of the other educt. This is why the selection of the pH conditions should be made with regard to Equation 18 (see Chapter 3), which gives the pH value for maximum peptide synthesis as a function of the respective ionization constants.

With reference to the previously mentioned synthesis of the dipeptide N^α-benzoylglycyl-glycine amide: if this reaction could take place in a mixture of water and a polyhydroxyalcohol (1:4) (v/v) and, furthermore, if the reaction temperature could be raised by 10°C, the free energy change of synthesis would arrive at negative values. That means, the equilibrium position would be over in the direction of proteosynthesis and the minimal requirements for the synthesis of peptides on a preparative scale would be fulfilled.

D. Influence of the Chain Length of the Reactants

Finally, it should be noted that both the acidity and the basicity of a peptide decreases with increasing chain length. Hence, the energy required for the proton transfer will also decrease; for example by approximately 2 to 3.5 kcal/mol when advancing from simple amino acids to di- and tripeptides.[7] Indeed, the equilibrium constants for the synthesis of central peptide bonds in proteins may be in the order of unity.[8] Consequently, the prospects of peptide synthesis will generally improve with growing chain length of the reactants, provided that the influence of ionizable side-chain functionalities can be neglected.

II. SHIFT OF CHEMICAL EQUILIBRIA

Besides the "manipulation" of ionic equilibria, there exist additional means to favor peptide synthesis at the expense of peptide hydrolysis. These aim at inducing a favorable shift of the chemical equilibria in protease-catalyzed reactions. All these expedients can be reduced formally to a single basic idea, that of energy coupling; a concept which is ubiquitous in living organisms. Thus, when synthesis of a peptide bond, for which

$$\Delta G^{o'}_{syn} > 0 \tag{3}$$

holds, is coupled to an energy-releasing reaction, for which

$$\Delta G_F^{o'} < 0 \tag{4}$$

holds, in such a way that the sum of the overall process is exergonic, then this can proceed spontaneously under suitable conditions since

$$\Delta G_\Sigma^{o'} = \Delta G_{syn}^{o'} + \Delta G_F^{o'} < 0 \tag{5}$$

In principle, the thermodynamically favored "auxiliary reaction" may be chemical or physical in nature and during enzymatic peptide syntheses both are used (*vide infra*). These alternatives share the capability of being able to continuously "sequestrate" the newly formed peptides, the concentrations of which are thereby kept well below their equilibrium concentrations in the reaction medium. This "extraction" process will proceed until, as a consequence of the continuously decreasing educt concentration, the product concentrations cannot be kept any longer below their equilibrium concentrations in the reaction medium.

In the following example, the dependence of the degree of synthesis, i.e., the relative amount of peptide synthesis, upon the initial concentration of the educts and upon the equilibrium position of both the overall process and its constituent reactions will be briefly described. Suppose an endergonic synthetic reaction which is assumed to be

$$A + B \rightleftharpoons C$$

the equilibrium relation of which reads

$$K_{syn}' = [C]/[A]^2 \tag{6}$$

(for the sake of simplicity the initial concentrations $[A_o]$ and $[B_o]$ of the educts were equalized) is followed by an exergonic reaction of the form

$$C \rightleftharpoons C^*$$

the equilibrium constant of which is given by the equation

$$K_F' = [C^*]/[C] \tag{7}$$

then the overall process becomes

$$A + B \rightleftharpoons C^*$$

and the total equilibrium constant can be expressed as

$$K_\Sigma' = [C^*]/[A]^2 \tag{8}$$

Then suppose, that the degree of synthesis, α, is defined as

$$\alpha = 1 - [A]/[A_o] \tag{9}$$

and that the material balance equation reads

$$[A_o] = [A] + [C] + [C^*] \tag{10}$$

with [A], [C], and [C*] being the equilibrium constants of A, C, and C*. Upon rearranging Equation 9 and bearing the relation of Equation 10 in mind one obtains

$$\alpha[A_o] = [C] + [C^*] \tag{11}$$

where $\alpha[A_o]$ is a measure of the absolute amount of synthesis.

By substituting terms from Equations 6 to 8, Equation 11 becomes

$$\alpha[A_o] = [A]^2(K'_{syn} + K'_{\Sigma}) \tag{12}$$

The rearrangement of Equation 9 to make A the subject followed by insertion into Equation 12 yields

$$\frac{\alpha}{(1 - \alpha)^2} = [A_o](K'_{syn} + K'_{\Sigma}) \tag{13}$$

Finally by solving Equation 13 for α one obtains

$$\alpha = \frac{2[A_o](K'_{syn} + K'_{\Sigma}) + 1 - \sqrt{4[A_o](K'_{syn} + K'_{\Sigma}) + 1}}{2[A_o](K'_{syn} + K'_{\Sigma})} \tag{14}$$

As can be inferred from the graphical illustration in Figure 1, Equation 14 states that at a given K'_{syn} the relative as well as the absolute amount of synthesis improves with increase in the initial concentrations of educts entering into the synthesis.

Evidently, the degree of synthesis is also crucially dependent upon the equilibrium position of the energetically favored "auxiliary" reaction which is defined by the equilibrium constant K_F' (cf. Equation 7). The greater K_F' (which itself is implicated in Equation 14 because $K_F' = K_{\Sigma}'/K'_{syn}$), the more strongly the overall reaction is driven toward the synthesis.

A. Solubility-Controlled Syntheses

The most frequently used technique to shift the overall equilibrium toward peptide synthesis by means of a favorable supplementary process takes advantage of the poor solubility in aqueous reaction media of fully protected peptides. In such a case, the overall equilibrium of the "coupled reaction", consisting of the actual peptide-bond-forming step and the subsequent precipitation of the resulting product, comprises not only the concentration of the educts and the product in its dissolved form but also the concentration of the product in its solid form. Suppose, that the partial equilibrium between the dissolved and the solid form of the product is far over in the direction of the latter, then the newly formed peptides are precipitated and thereby continuously removed from solution. This process will proceed until the concentration of one educt is so reduced that the equilibrium concentration of the dissolved peptides equals their solubility in the reaction medium. The degree of synthesis, α, for reactions the product, but not the educts of which are precipitated, can be obtained by solving Equation 6 for [A] which is then inserted into Equation 9 to give

$$\alpha = 1 - \sqrt{[C_S]/K'_{syn}[A_o]} \tag{15}$$

where $[A_o]$ is again the initial concentration of the two educts, and where [C] was replaced by the maximal product solubility in the respective reaction medium, i.e., by the solubility $[C_S]$. To further reduce the solubility of the prospective peptides in a given buffer solution, the hydrophobicity of the educts can be increased both by the introduction of suitable, apolar

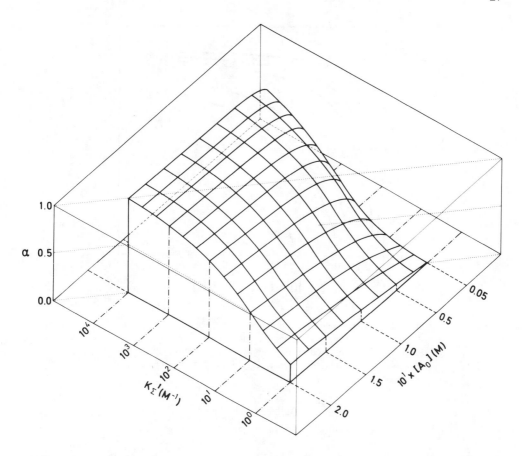

FIGURE 1.   Peptide synthesis: The influence of $[A_0]$, the initial concentration of the educt, and K'_Σ, the overall equilibrium constant, on α, the relative degree of synthesis (with logarithmic K'_Σ scale). The curves of the 3°-plot were calculated from Equation 14 with $K'_{syn} = 0.5\ M^{-1}$.

protector groups and by enhancing the ionic strength of the solution through the addition of salts, i.e., one takes advantage of so-called ''salting-out'' effects. (In doing so, care must of course be taken to ensure that the educts remain in solution.) The method of ''precipitation-mediated synthesis'' was originally applied by Bergmann and his collaborators during their first enzymatic syntheses[9,10] and has subsequently proved to be an efficient approach to enzymatic peptide chemistry. The insoluble peptidic products are conveniently purified and, as illustrated in Figure 2, the product yield may be almost quantitative when the starting concentration of the educts is sufficiently high and the solubility of the product is sufficiently low; or in other words, if a reasonable solubility differential between the prospective peptide and its educts can be arranged.

This proviso also holds true for another solubility-controlled technique which is also based on the removal from equilibrium of the newly generated products. This procedure, however, aims primarily not at the precipitation from the solution of the peptidic products but rather at their transfer from an aqueous to a water-immiscible organic layer of a biphasic reaction system.[11-13] Besides the already discussed merits of organic co-solvents, the free energy change of this transfer provides the driving force for peptide synthesis.

The degree of synthesis is a function of the initial concentration of the educts (for simplicity each is taken to have the same concentration), of the volume ratio of the phases, and of the partition coefficients of the reactants and is described by the following equation, which represents a modified version of the relation originally developed by Martinek et al.[12]

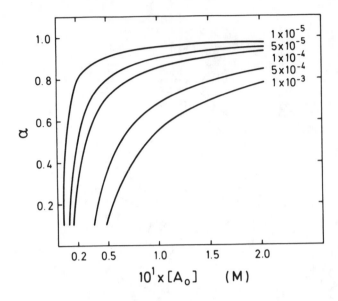

FIGURE 2.   Dependence of the extent of peptide synthesis, α, upon the solubility, $[C_S]$, of the product and the initial concentration, $[A_o]$, of the educts. The different fixed molar solubilities are indicated at the ends of the curves. The equilibrium constant, K'_{syn}, was $0.5\ M^{-1}$; the curves were calculated from Equation 15.

$$\frac{\alpha}{(1\ -\ \alpha)^2} = \frac{[A_o]K_W(1\ +\ \beta P_C)(1\ +\ \beta)}{55.5(1\ +\ \beta P_A)(1\ +\ \beta P_B)} \qquad (16)$$

where K_W is the equilibrium constant in the aqueous layer and $[A_o]$ is the total starting concentration of A referred to the overall volume of the biphasic system. The volume ratio of the organic to the aqueous phase is given by β, while the partition coefficients of the educts A and B, and the product C between the organic and the aqueous phase are represented by $P_A = [A]_{org}/[A]_W$, $P_B = [B]_{org}/[B]_W$ and $P_C = [C]_{org}/[C]_W$, respectively. As stated in Equation 16 and illustrated in Figure 3, the prospects for peptide synthesis can be improved by lowering the percentage of water in the whole system, provided that the reaction product is more readily extracted from the aqueous phase into the organic phase than the educts. The larger the difference between the partition coefficients of the product, P_C, and of the educts, P_A and P_B, the more valid this becomes. Although peptide bond formation in biphasic systems benefits *a priori* from the pronounced hydrophobicity of the products, which in general exceeds that of educts possessing either free α-carboxyl or free α-amino groups, the success of a peptide synthetic process largely depends upon the judicious choice of suitable solvent systems and the reasonable use of protector groups.

If, in addition to the above, the chemical nature of either the organic layer or the newly formed peptide favors the precipitation of the latter, then a "third" phase — the solid phase — will come into being. The free energy change of this "phase transfer" will provide an additional shift toward peptide bond synthesis.

B. Molecular Traps

In a figurative sense, the precipitation of insoluble peptides can be considered as a "product-trapping" method. An alternative technique that has been used in preference during protein semisyntheses (see Chapter 10) profits from the presence of a specific molecular trap capable of specifically scavenging the desired products. In practice, a newly synthesized

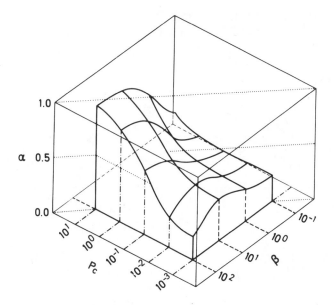

FIGURE 3. Peptide synthesis in biphasic aqueous-organic systems. The relative amount of synthesis, α, is plotted as a function of β, the volume ratio of the organic and aqueous phase, and P_C, the partition coefficient of the product (with logarithmic β and P_C scale). The curves of the 3°-representation are calculated from Equation 16 with $K_W = 10\ M^{-1}$, $[A_o] = 2\ M$, and $P_A = P_B = 0.1$.

peptide is noncovalently bound (trapped) by another peptide or protein to form a stable complex. This approach does not basically differ in principle from previously mentioned methods since the energy-releasing process of product chelation can significantly overcompensate for the energy-consuming process of peptide bond formation. The "trapping agents" need not be exclusively peptidic in nature, for example as protease inhibitors, antibodies, or receptors and the like. In fact, any substance capable of forming an intra- or intermolecular complex with the reaction product, but not with the educts, may act as a scavenger.[14]

C. Effects of Concentration

The formation of a peptide bond represents a condensation process during which two molecules are connected to each other via an amide bond to generate a new molecule. The product yield depends not only upon the equilibrium of this reaction but also upon the initial concentration of the educts. The degree of synthesis, α, i.e., the relative amount of synthesis of an arbitrary dipeptide of the form A-B, from the constituent amino acids A and B (which contribute the α-imino and the α-carbonyl group, respectively, to the prospective amide bond) can be deduced from the law of mass action in the following way; suppose the initial concentrations of the amino acids A and B are $[A_o]$ and $[B_o]$, respectively. Then, under equilibrium conditions the concentration of the product, the dipeptide A-B, becomes $\alpha[A_o]$, while the concentrations of the educts A and B are reduced to $[A_o] - \alpha[A_o]$ and $[B_o] - \alpha[A_o]$, respectively. Hence, the equilibrium relation reads

$$\frac{\alpha[A_o]}{([A_o] - \alpha[A_o])([B_o] - \alpha[A_o])} = K'_{syn} \qquad (17)$$

where K'_{syn} is the equilibrium constant for peptide synthesis. Solving this equation explicitly

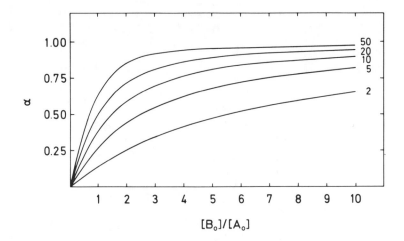

FIGURE 4. Variation with the initial concentration of the educts, $[A_o]$ and $[B_o]$, of α, the degree of synthesis. The curves are calculated on the basis of Equation 18 with $K'_{syn} = 100\ M^{-1}$. The initial millimolar concentrations of A were held constant at the values indicated at the end of the curves.

for α gives us the degree of synthesis

$$\alpha = \frac{K'_{syn}([A_o] + [B_o]) + 1}{2K'_{syn}[A_o]}$$
$$- \frac{\sqrt{(K'_{syn})^2([A_o] - [B_o])^2 + 2K'_{syn}[A_o] + [B_o] + 1}}{2K'_{syn}[A_o]} \qquad (18)$$

the concentration dependence of α at a given K'_{syn} is illustrated in Figure 4 which shows that the relative amount of synthesis can be significantly increased by increasing the initial concentration of the respective educts. It is not all that rare during peptide synthetic chemistry to find that $[A_o] \neq [B_o]$. In particular during enzymatic peptide synthesis, it is frequently the case (see below) that the concentration of the amine component exceeds that of the carboxyl component; i.e., $[A_o] < [B_o]$. Therefore, the effect of the concentration ratio of A to B on the degree of synthesis can be of the utmost importance. As can be inferred from Figure 4, the product yields, relative to A, can be remarkably improved by increasing the initial concentration of B.

The synthesis of a peptide bond is normally accompanied by the release of a water molecule. The influence of the water concentration on the degree of synthesis was not explicitly taken into account during the previous considerations. However, it is an integral part of the equilibrium constant K'_{syn}. The presence of organic cosolvent causes, in addition to the already mentioned shift of the ionic equilibrium, a drastical reduction of the water concentration and, consequently according to the law of mass action, a shift of the chemical equilibrium in favor of peptide synthesis.

D. Temperature-Dependent Equilibrium Shifts
The temperature dependence of the equilibrium position in a peptide synthetic reaction as defined by the van't Hoff equation

$$\frac{d \ln K}{d(1/T)} = -\frac{\Delta H^o}{R} \qquad (19)$$

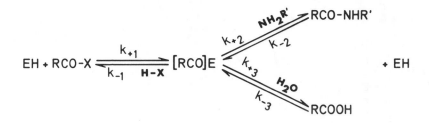

FIGURE 5.   Aminolysis and hydrolysis in a protease-mediated reaction

can also be used to influence the outcome of the peptide bond-forming step. According to Le Chatelier's principle, a reaction will proceed in the direction in which heat is absorbed by the equilibrium reaction system. As a consequence, the endothermic — ($\Delta H^{\circ\prime}_{syn} > 0$) — process of peptide bond formation[15] will be enhanced with rising temperature. For example, the synthesis of protected dipeptides was observed to proceed exergonically — ($\Delta G^{\circ\prime}_{syn} < 0$) — at temperatures above 60°C.[16]

III. NONEQUILIBRIUM APPROACHES: KINETICALLY CONTROLLED SYNTHESES

In the course of any thermodynamically controlled process those reagents with the lowest free energy will finally accumulate. In protease-catalyzed reactions the cleavage products of the general formula $RCOO^-$ and $^+NH_3R'$ are usually more stable than the "intact" peptides of the general formula RCO-NHR'. Consequently, the concentration of $RCOO^-$ and $^+NH_3R'$ will exceed that of the RCO-NHR' as soon as an equilibrium state has been attained. From the point of view of peptide synthesis this is a highly undesirable situation and to prevent this from occurring peptide bond formation is commonly coupled to an energy-releasing process, as described above.

However, this procedure is by no means a *conditio sine qua non* for successful peptide synthesis. Notwithstanding a thermodynamic caveat, appreciable product yields may be obtained during "uncoupled" peptide synthesis if the reaction is terminated well before the equilibrium is established. In this case, the products which accumulate are those that are produced most rapidly and destroyed most slowly;[17] i.e., the reaction is controlled kinetically rather than thermodynamically. This is illustrated by the scheme of a protease-catalyzed reaction (Figure 5), in which a transient acyl-enzyme complex [RCO]E occurs (cf. Chapter 4) and is competitively deacylated via aminolysis or hydrolysis.

Provided, that the deacylation event is the rate-determining step of the catalytic process, and if

$$k_{+2}[NH_2R'] > k_{+3}[H_2O] \qquad (20)$$

holds true, where k_{+2} and k_{+3} are the rate constants for the aminolytic and hydrolytic cleavage of the acyl-enzyme complex, then the concentration of the product RCO-NHR' of the protease-catalyzed reaction temporarily exceeds its equilibrium concentration. (The possibility, that the inequality [Equation 20] actually proves right, is favored by the observations of Fersht et al.,[17] who determined the rate constants for the attack of amines on various acyl-chymotrypsin complexes.)

When the educts and the product are dissolved in the reaction medium, the dependence of the product yield, P, upon the concentration of the amine NH_2R' (being present in a molar excess) and the ratio of the rate constants, k_{+2}/k_{+3}, is approximated by the following relation:[18]

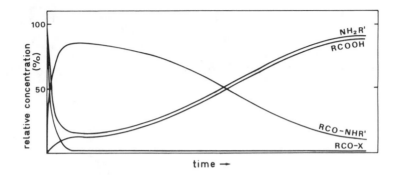

FIGURE 6.   Variation with time of the educt and product concentrations in a protease-catalyzed reaction, where k_{+2} [NH$_2$R'] > k_{+3} [H$_2$O].

$$P(\%) = \frac{100[NH_2R']}{[NH_2R'] + [H_2O]k_{+3}/k_{+2}} \qquad (21)$$

(A more detailed calculus of the product yield has recently been presented by Könnecke et al.[19]) It can be inferred from Equation 21 that the synthetic yield is improved by increasing the concentration of the amine (cf. this chapter, Section C). Peptide bond formation can be further promoted by strengthening the "deacylating potential" of amine — as expressed by the rate constant, k_{+2}, of aminolysis — toward the acyl-enzyme complex. (This can be achieved for instance by suitably protecting the α-carboxyl group of the amine.)

In the course of a protease-catalyzed process, the target product RCO-NHR' merely represents an intermediate form which is subject to subsequent secondary-hydrolysis. The concentration of RCO-NHR' is a function of time and possesses a maximum (cf. Figure 6), where the yield of the "intermediate" may significantly exceed the equilibrium level. It is therefore of utmost importance to stop the reaction at the "kinetic" optimum.

IV. SOME ADDITIONAL CONSIDERATIONS DUE TO THE PROTEASES THEMSELVES

Up to this point, we have discussed the thermodynamic or kinetic aspects of peptide synthesis without reference to the actual "mediators" of the synthetic reactions, i.e., the proteases. This may be an inadmissible over simplification since in some instances protease-immanent features are at variance with an otherwise favorable expedient. Thus the addition of organic solvents to a reaction mixture or an elevation of temperature, which on thermodynamic criteria should favor peptide bond formation, may actually result in a considerable reduction in activity of the biological catalyst. Furthermore, the choice of suitable protection groups depends not only on their capacities to lower the acidity or basicity of a given educt or to promote the insolubility of a prospective product, but must also account for favorable interactions between substrates and enzyme. Last but not least, the thermodynamically optimal pH value as defined by Equation 18 (see Chapter 3), should be largely compatible with the pH optimum of the respective protease. However, despite these reservations, protease-immanent properties offer considerable scope for peptide synthetic purposes. The catalytic action of many proteases is not confined exclusively to peptide bonds but also extends to ester- and amide bonds. The utilization of these additional activities to synthesize a peptide bond involves the substitution of the ester- or amide group by an amino-acid or peptide derivative. The first step of such a reaction sequence, i.e., the removal of the ester- or amide group, represents a highly exergonic process that can drive the subsequent step, i.e., formation of the peptide bond itself. Apart from the order of succession, this procedure is reminiscent of the previously mentioned strategy of coupling an energy-consuming to an energy-releasing reaction (cf. this chapter, Section II). Although in the present case the exergonic process precedes the endergonic one, the esterified or amidated substrates may be considered as being in a "pre-activated" state. In addition, it has been shown that the esterase and peptidase activities are optimally expressed at different pH values. Of course, while these properties may be favorably exploited to suppress an eventual a posteriori hydrolysis of the newly formed peptide, this nevertheless further complicates the choice of a suitable pH for the reaction.

In conclusion, it is often a delicate matter to balance the bright and the dark side of the picture. As the case may be, a compromise has to be found that takes regard of both physico-chemical requirements and protease-inherent peculiarities. [*Note:* It is indeed remarkable, that the proteases possessing hydrolytic as well as aminolytic and sometimes esterolytic activities, actually bear a close analogy to the activities normally participating in the process of protein biosynthesis. Thus, the so-called peptidyl transferase, a ribosomal protein, is also considered to possess hydrolytic, aminolytic, and esterolytic capacities.[20] Furthermore, the molecular mechanisms, which are considered to govern the formation of peptide bonds during both ribosomal and the nonribosomal protein biosynthesis[21,22] as well as the catalysis through serine[23] and cysteine proteases[24], impressively parallel each other (Figure 7).]

In view of these functional and mechanistic analogies, the term "process-evolution" coined by F. Lipmann with reference to the biosynthesis of peptides[25] appears to be also applicable to the family of enzymes, whose activities are directed toward peptide bonds. Thus, the choice between a proteosynthetic or proteolytic action for a given enzyme will depend primarily upon the energetics of the reactions to be catalyzed rather than upon the molecular mode of action of the enzyme. Certainly in the case where a protease-catalyzed reaction is allowed to proceed in aqueous solution under physiological conditions then, as commonly occurs for in vivo protease-controlled processes, proteolysis will largely predominate. However, in the more polar environment which may prevail during ribosome mediated in vivo synthesis of proteins and peptides, proteosynthesis should be favored. Consequently, with a knowledge of the similarities between the mechanisms of peptide bond synthesis and degradation, the original concept of a protease-catalyzed protein biosynthesis becomes less incongruous than might have appeared when the first details of ribosomal protein biosynthesis became known.

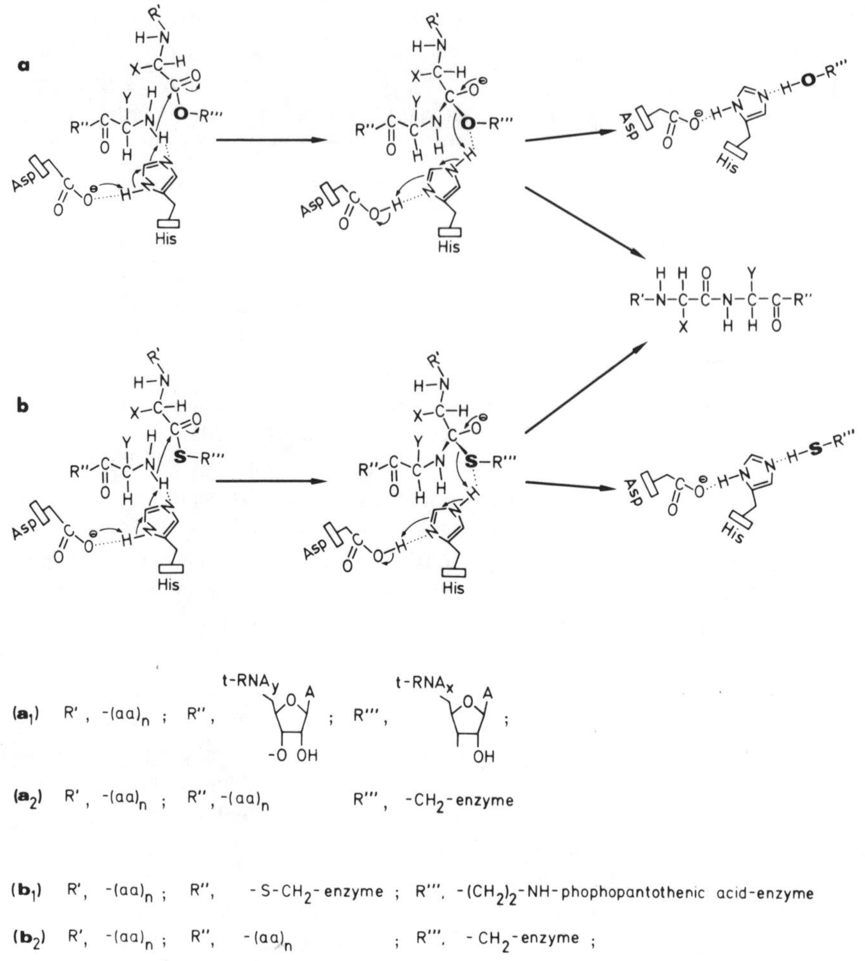

FIGURE 7. Synthesis of peptide bonds, (a_1) ribosomal, as catalyzed by the peptidyltransferase;[21] (b_1) nonribosomal, as catalyzed by a peptide antibiotic synthetase;[22] (a_2) via chymotryptic catalysis; (b_2) via papain-catalysis. (the proposed reaction mechanisms of the protease-catalyzed syntheses imply an aminolytic rather than the more common hydrolytic deacylation of the acyl-enzyme complex. As far as the protease papain is concerned, the aspartic acid residue is substituted by an asparagine residue.[24] X and Y designate amino acid side-chains as well as the corresponding t-RNAs, whereas A represents the nitrogen base adenine).

REFERENCES

1. **Carpenter, F. H.,** The free energy change in hydrolytic reactions: The non-ionized compound convention, *J. Am. Chem. Soc.,* 82, 1111, 1960.
2. **Michaelis, L. and Mizutani, M.,** Die Dissoziation der schwachen Elektrolyte in wässrig-alkoholischen Lösungen, *Z. Phys. Chem. (Leipzig),* 116, 135, 1925.
3. **Mizutani, M.,** Die Dissoziation der schwachen Elektrolyte in wässrig-alkoholischen Lösungen. Die Beziehungen zwischen chemischer Konstitution und Alkoholempfindlichkeit der Säuren und Basen, *Z. Phys. Chem.(Leipzig),* 116, 350, 1925.
4. **Homandberg, G. A., Mattis, J. A., and Laskovski, M., Jr.,** Synthesis of peptide bonds by proteinases. Addition of cosolvents shifts peptide bond equilibria toward synthesis, *Biochemistry,* 17, 5220, 1978.

5. **Richardson, G. M.,** The principle of formaldehyde, alcohol, and acetone titrations. With a discussion of the proof and implication of the zwitterionic conception, *Proc. R. Soc. London B,* 115, 121, 1934.

6. **Perrin, D. D.,** The effect of temperature on pK values of organic bases, *Aus. J. Chem.,* 17, 484, 1964.

7. **Jakubke, H.-D. and Kuhl, P.,** Proteasen als Biokatalysatoren für die Peptidsynthese, *Pharmazie,* 37, 89, 1982.

8. **Laskowski, M., Jr.,** The use of proteolytic enzymes for the synthesis of specific peptide bonds in globular proteins, in *Semisynthetic Peptides and Proteins,* Offord, R. E. and DiBello, C., Eds., Academic Press, New York, 1978, 255.

9. **Bergmann, M. and Fraenkel-Conrat, H.,** The enzymatic synthesis of peptide bonds, *J. Biol. Chem.,* 124, 1, 1938.

10. **Bergmann, M. and Fruton, J. S.,** Some synthetic and hydrolytic experiments with chymotrypsin, *J. Biol. Chem.,* 124, 321, 1938.

11. **Semenov, A. N., Berezin, J. V., and Martinek, K.,** Peptide synthesis enzymatically catalyzed in a biphasic system: water-water-immiscible organic solvent, *Biotechnol. Bioeng.,* 23, 355, 1981.

12. **Martinek, K., Semenov, A. N., and Berezin, J. V.,** Enzymatic synthesis in biphasic aqueous-organic systems. I. Chemical equilibrium shift, *Biochem. Biophys. Acta,* 658, 76, 1981.

13. **Martinek, K. and Semenov, A. N.,** Enzymatic synthesis in biphasic aqueous-organic systems. II. Shift of ionic equilibria, *Biochim. Acta,* 658, 90, 1981.

14. **Chaiken, J. M.,** Semisynthetic peptides and proteins, *Crit. Rev. Biochem.,* 11, 255, 1981.

15. **Borsook, H.,** Peptide bond formation, *Adv. Prot. Chem.,* 8, 127, 1953.

16. **Flegmann, A. W. and Tattersall, R.,** Energetics of peptide bond formation at elevated temperatures, *J. Mol. Evol.,* 12, 349, 1978.

17. **Petkov, D. D.,** Enzyme peptide synthesis and semisynthesis: Kinetic and thermodynamic aspects, *J. Theor. Biol.,* 98, 419, 1982.

18. **Fersht, A. R., Blow, D. M., and Fastrez, J.,** Leaving group specificity in the chymotrypsin-catalyzed hydrolysis of peptides. A stereochemical interpretation, *Biochemistry,* 12, 2035, 1973.

19. **Petkov, D. D. and Stoineva, I. B.,** Nucleophile specificity in chymotrypsin peptide synthesis, *Biochem. Biophys. Res. Commun.,* 118, 317, 1984.

20. **Könnecke, A., Schellenberger, V., Hofmann, H.-J., and Jakubke, H.-D.,** Die Partitionskonstante als Effizienzparameter von Nucleophilen bei enzymkatalysierten kinetisch kontrollierten Peptidsynthesen, *Pharmazie,* 39, 785, 1984.

21. **Weissbach, H. and Pestka, S.,** *Molecular Mechanisms of Protein Biosynthesis,* Academic Press, New York, 1977.

22. **Nierhaus, K. H., Schulze, H., and Cooperman, B. S.,** Molecular mechanisms of the ribosomal peptidyltransferase center, *Biochem. Int.,* 1, 185, 1980.

23. **Kleinkauf, H., and van Döhren, H., Eds.,** A survey of enzymatic peptide formation, in *Peptide Antibiotics-Biosynthesis and Functions,* Walter de Gruyter, Berlin, 1982, 3.

24. **Blow, D. M.,** Structure and mechanisms of chymotrypsin, *Acc. Chem. Res.,* 9, 145, 1976.

25. **Drenth, J.,** Proteolytic enzymes. General features of their mode of action, *Recl. Trav. Chim. Pays-Bas,* 99, 185, 1980.

26. **Lipmann, F.,** Attempts to map a process evolution of peptide biosynthesis, *Science,* 173, 875, 1971.

Chapter 6

ADVANTAGES OF ENZYMATIC PEPTIDE SYNTHESIS

ABSTRACT

In view of the extensive efforts required to improve the anything but favorable prospects of protease-catalyzed peptide bond formation, it is worthwile considering the benefits resulting from adopting the enzymatic approach to peptide synthetic chemistry.

I. SPECIFICITY

The most obvious advantage of enzymatic syntheses results from one of the most impressive biological phenomena; namely, the pronounced capacity of an enzyme to catalyze chemical reactions with an otherwise unattainable specificity. Like most other enzymes, the proteases tend to restrict their activities to one or to a very limited number of closely related substrates. As a consequence, the use of proteases as catalysts for the peptide-bond-forming steps of in vitro syntheses offers ideal opportunities to suppress most of the undesired side reactions often encountered during relatively unspecific chemical syntheses.[1-3]

One can distinguish between three levels of specificity, each of which individually contributes to the merits of enzymatic peptide synthesis.

A. Structural Specificity

Structural specificity enables the proteases to select from a mixture of compounds only authentic substrates possessing distinct structural features. One or more iminoacyl subunits of the type (−NH−CHR−CO−) usually represent the criterion for a general substrate, whereas the individual, structural specificity is embodied by the preference of a given protease for distinct amino acid side chains (R) which are located in the immediate neighborhood of the sensitive peptide bond. Presently, it is common usage to discriminate between the primary and the secondary specificity.[4] The primary specificity is mostly determined by a strong bias of the protease in favor of a single amino acid which can sometimes be replaced by different though structurally related amino acids; however, the position of these within the respective substrate is unequivocally defined. Conversely, the secondary specificity comprises a more extended area of the substrate, which may involve amino acids being placed on both sides of the protease-sensitive peptide bond. The preference of the protease for the amino acid residues occupying these positions is significantly weaker compared to that observed for the primary specificity. The presence of the correct amino acid, as dictated by the primary specificity of a protease, is a mandatory prerequisite for successful recognition and catalysis by the enzyme. However, satisfaction of the secondary specificity is merely faculative, although secondary enzyme-substrate interactions may considerably influence the binding affinity as well as the catalytic efficiency.

Due to the specificity of the reactions catalyzed by the proteases and hence the relative lack of undesired by-products, the time-consuming and laborious purification steps which are often required during the chemosyntheses of peptides can be considerably simplified or even omitted. Scarcely any of the individual reactions performed during the chemical synthesis of a peptide will result in a 100% yield of product. Usually residual unreacted educts are left behind and, in addition, by-products, the physical and chemical nature of which differ only slightly from that of the main product, are created. It is an indispensable "must" to remove these by-products prior to embarking upon the next stage of the synthetic pathway in order to avoid the synthesis of a further variety of novel by-products. By contrast, given the specificity of protease-catalyzed reactions where only one peptide in a product mixture might exhibit the structural features identifying it as a substrate of a given protease, only this peptide will participate in the subsequent reaction. In fact, the original impurities are not eliminated by this procedure, but the newly formed product is likely to exhibit physical and chemical properties widely differing from those of the original by-products. At this stage of synthesis, therefore, the latter may be rather conveniently separated from the novel main product.

B. Regiospecificity

The regiospecificity of the proteases is associated with a strictly limited portion of the substrate; namely, with the α-carboxyl and the α-imino group of the sensitive peptide bond. Apart from these two groups, all the other potentially reactive groups of the substrate are not involved in the protease-controlled catalyses.

The impact of the regiospecific action of the proteases may be illustrated by the fact that 13 out of the 20 "codogenous" amino acids possess a third- or side-chain functionality. The integrity of these potentially reactive groups is maintained during enzymatic syntheses, however, unless suitably protected, they are likely to give rise to undesired side reactions during chemical syntheses. In the presence of the commonly used relatively unspecific chemical coupling agents, the semipermanent side-chain protection is obligatory as far as aspartic acid, glutamic acid, lysine, and cysteine residues are concerned. The incorporation of protector groups is conditionally obligatory, and is actually carried out as a rule, in the case of threonine, serine, tyrosine, arginine, and histidine residues. Conversely, the blocking of the third functionalities of glutamine, asparagine, tryptophan, and methionine residues is merely a matter of personal choice. Although the major difficulties in chemical syntheses can be overcome by the introduction of protecting groups, some problems still remain. Thus, in some instances the shielding of the functional groups is not absolute, i.e., a residual activity may persist, a shortcoming often observed with the guanidine function of arginine (see below). Furthermore, the reiterated removal of the temporary α-amino blocking groups prior to each elongation of the growing peptide chain does not always proceed selectively; that is to say, with increasing chain length the semipermanent protecting groups may also be attacked. Finally, after the completion of the synthesis the removal of the semipermanent blocking groups may also cause undesired side reactions resulting in chemically heterogeneous end-products (a more detailed description of the chemistry of protecting groups can be found in References 1 to 3. In view of these considerations, the advantages resulting from the exploitation of protease-inherent regiospecificity rather than chemical blocking to preserve the identity of side-chain functionalities during peptide synthesis cannot be overestimated.

C. Stereospecificity

The proteases display a stereospecificity of substrate recognition which is determined by configurational patterns of the substrates. As far as amino acid residues having an asymmetric α-carbon atom are concerned, the proteases will act almost exclusively upon a single optical isomer while ignoring its optical antipode. With regard to the primary specificity, proteases usually demonstrate an absolute preference for the L-enantiomeric form of given amino acid residue. However, as far as the secondary specificity is concerned, it may indeed happen that both optical isomers are accepted as substrates; although amino acids having an L-configuration are still markedly preferred (see below).

This property of stereospecificity is turned to considerable — and presumably the most prominent — advantage during enzyme-catalyzed peptide synthesis. Since, as noted above, most proteases demonstrate a strong preference for the L-enantiomer of an amino acid they can be used to resolve heterochiral mixtures of a racemic substrate. In a peptide bond forming reaction, therefore, the ability of proteases to exert stereospecific control in their catalysis

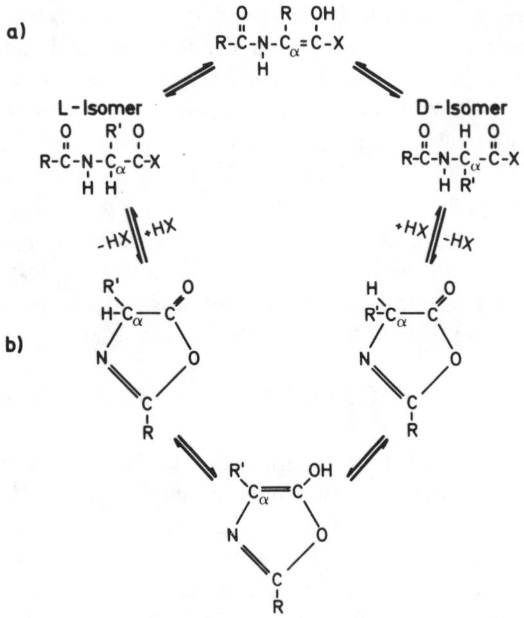

FIGURE 1. Mechanisms of racemization: Inversion of configuration at the C^α-atom by (a) enolization or (b) by oxazolone formation.

results in optically pure homochiral products, even though the starting material might have originally been an optically heterogeneous mixture.

The most useful feature of the enzymatic approach to peptide synthetic chemistry is certainly its ability to provide — due to the stereospecific action of the proteases — for an outstanding way of maintaining the chirality or handedness of the reactants. This advantage cannot be overestimated, since racemization has been a major problem in the field of chemical peptide synthesis throughout its history. Consequently, Gross and Meienhofer hit the nail on the head when they stated: "The preservation of chiral integrity is a cardinal criterion for judging the relative merits of synthetic methods".[5]

During chemical peptide-bond formation both the carboxyl activation and the subsequent coupling step may proceed in a sterically unspecific manner and as a consequence, give rise to the generation of optically heterogeneous by-products. Two alternative pathways may lead to a configurational inversion at the α-carbon atom of the carboxyl component (Figure 1) (a) via simple enolization or (b) through the formation of 5(4H)-oxazolone.[1,6] Compounds that can be inverted to their respective mirror image isomer by both mechanisms are at least two orders of magnitude more susceptible to the 5(4H)-oxazolone-mediated pathway than to enolization.[6]

Although the oxazolone-mediated hazards of racemization can be minimized by incorporating N^α-urethane protected amino acids (for which the less rapid enolization pathway is well established) during a stepwise elongation of the growing peptide chains,[6] the 5(4H)-oxazolone pathway remains relevant as far as racemization of both the penultimate[7,8] and the ultimate amino acid residue of N^α-blocked peptide fragments is concerned. The verdict of Kemp[6] on the current situation in chemical peptide synthesis illustrates the point: "...despite much planning and efforts the problem of racemization remains most acute for the amide-forming steps of synthesis by fragment condensation".

Both the enolization and the oxazolone-forming mechanism of C^α-inversion are initiated by a base-catalyzed reversible withdrawal of the C^α-hydrogen, resulting in symmetric intermediates which can be unspecifically reconverted to L- and D-enantiomers (Figure 1). Similarly, the amino acid racemase mechanism also leads via C^α-hydrogen-abstraction to a symmetric intermediate, which is bound through a Schiff's base to pyridoxal phosphate,[9] the co-factor of the racemase. Although the chemistry of α-carboxyl activation, via formation of active esters, is comparable in conventional and enzymic peptide synthesis, the relative three-dimensional structure of the binding-site of the proteases obviously enables catalysis to be stereospecific thus preventing the proteases from acting as racemases.

From these considerations it can be concluded that the preparation of peptidic chirons,[10] chiral peptidic synthons, with the desired asymmetric centers is best accomplished enzymatically. In fact this conclusion holds true not only for peptide synthetic reactions, but it also applies in a more general sense to many organic syntheses.

II. OTHER POSITIVE ASPECTS OF ENZYMATIC PEPTIDE SYNTHESIS

Enzymes are not only nature's most specific catalysts, but they are also characterized by their capacity to accelerate chemical reactions to a unique degree. Thus, the reaction rate of protease-catalyzed peptide bond formation significantly exceeds that usually observed for conventional procedures, despite the enzymes being present in relatively low concentrations. In this context, the opportunity to reuse proteases repeatedly offers an additional advantage, particularly, if the proteases are used in an immobilized state. Furthermore, protease-mediated syntheses, in common with other enzyme-based technologies, are carried out under relatively mild conditions thereby preventing the hazards eventually associated with chemical procedures.

REFERENCES

1. **Wünsch, E.,** Synthese von Peptiden, in *Houben-Weyl, Methoden der organischen Chemie,* Vol. 15, Parts, 1 and 2, G. Thieme Verlag, Stuttgart, 1974.
2. **Gross, E. and Meienhofer, J.,** *The Peptides: Analysis, Synthesis, Biology,* Vols. 1-5, Academic Press, New York, 1979, 183.
3. **Bodanszky, M. and Martinek, J.,** Side-reactions in peptide synthesis, *Synthesis,* 333, 1981.
4. **Fruton, J. S.,** Proteinase-catalyzed synthesis of peptide bonds, *Adv. Enzymol. Relat. Areas Mol. Biol.,* 53, 239, 1982.
5. **Gross, E. and Meienhofer, J.,** The peptide bond, in *The Peptides: Analysis, Synthesis, Biology,* Vol. I, Gross, E. and Meienhofer, J., Eds., Academic Press, New York, 1979, 1.
6. **Kemp, D. S.,** Racemization in peptide synthesis, in The *Peptides: Analysis, Synthesis, Biology,* Vol. I, Gross, E. and Meienhofer, J., Eds., Academic Press, New York, 1979, 315.
7. **Weygand, F., Prox, A., and König, W.,** Racemisierung der vorletzten carboxylendständigen Aminosäure bei Peptidsynthesen, *Chem. Ber.,* 99, 1446, 1966.
8. **Dzieduszyka, M., Smulkowsky, M., and Taszner, E.,** Racemization of amino acid residue penultimate to C-terminal amino acid, *Pol. J. Chem.,* 53, 1095, 1979.
9. **Dixon, M. and Webb, E.,** Enzyme mechanisms, Enzyme cofactors, in *Enzymes,* 3rd ed., Longman Group, London, 320, 508, 1979.
10. **Hanessian, S.,** *Total Synthesis of Natural Products: The "Chiron" Approach,* Pergamon Press, Oxford, 1983.

Chapter 7

PROTEASES AS BIOCATALYSTS FOR THE SYNTHESIS OF MODEL PEPTIDES

I. INTRODUCTION

Over the years, research on the optimization of organic syntheses by the application of enzymes has been the subject of considerable interest.[1,2] In 1949, a report on "state-of-the-art" chemical synthesis of peptides was presented by J. S. Fruton. On this occasion the author also pointed out the possibility of applying proteases to preparative scale peptide synthesis.[3]

In the 1950s a series of studies were performed, in particular by Fruton and his collaborators, which dealt with enzymatic peptide synthesis via transamidation. (Studies on this topic are discussed in detail in Reference 4). However, those investigations were stimulated more by the idea of a possible participation of proteases in the pathway of protein biosynthesis rather than by any intention to synthesize peptides on a preparative scale. Certainly, this was neither the primary aim of Wieland, Determann, and co-workers when in the 1960s they tried to elucidate the molecular mechanisms of the plastein reaction (*vide infra*, Chapter 9).

Evidently, the interest in protease-catalyzed peptide syntheses markedly waned as the current concept of ribosome mediated protein biosynthesis developed. Fruton's previous suggestion concerning protease-controlled peptide synthesis was only taken up in the late 1970s when the Japanese groups of Isowa and Morihara initiated their pioneering studies on the ability of a series of proteases to act as catalysts in peptide synthetic chemistry. Because of their importance in establishing enzymatic peptide synthesis, these and additional model studies are presented here in some detail. For the sake of clarity, the discussion of these experiments follows the order in which individual protease classes are harnessed in the catalysis of peptide bond formation. The first to be described will be the family of serine proteases, which are characterized by a highly reactive serine residue in the active site, the hydroxyl function of which becomes acylated upon the formation of a covalent substrate-enzyme complex.

II. α-CHYMOTRYPSIN

The endopeptidase α-chymotrypsin preferentially acts upon peptide bonds adjacent to the carbonyl groups of aromatic amino acid residues, such as phenylalanine, tyrosine, or tryptophan.[5,6] Furthermore, peptides having histinine, leucine,[7] methionine, or threonine residues,[8,9] in the P_1-position (shorthand notation as proposed by Schechter and Berger[10]; cf. Figure 1) are also recognized as substrate, although less avidly. The primary specificity of α-chymotrypsin which is definitely associated with the P_1-position may be extended if, in addition to the above-mentioned peptidase activities, the esterase activity of the enzyme is also considered.[11] The secondary specificity of the enzyme is expressed to a minor degree, so that there exists a certain preference for hydrophobic amino acid residues in the P_2- and P_1'-position of the chymotryptic substrates.[12]

Pioneering work on chymotrypsin-mediated peptide bond formation was reported by Bergmann and Fruton as early as 1938 (see Chapter 2).[6] The authors succeeded in preparing the dipeptide Bz-Tyr-Gly-NHPh starting from Bz-Tyr-OH and H-Gly-NHPh. Four decades later, α-chymotrypsin-catalyzed proteosynthesis was systematically investigated by Morihara and Oka.[13] In these studies Nα-protected aromatic amino acid and peptide ethyl esters served as acyl group donors (carboxyl components) whereas a variety of amino acid derivatives,

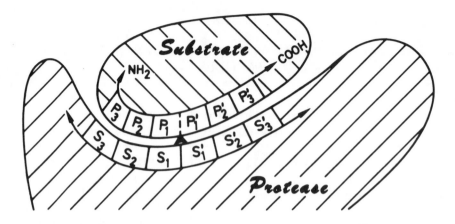

FIGURE 1.   Schematic representation of substrate-protease interactions. (From Schechter, I. and Berger, A., *Biochem. Biophys. Res. Commun.*, 27, 157, 1967. With permission.)

peptides, and their derivatives were used as acceptor nucleophiles (amine components). The data obtained from the enzymatic synthesis of a series of model peptides enabled the authors to predict the conditions most suitable to giving optimal yields. In contrast to chymotryptic proteolysis, rather high concentrations of the protease and a more alkaline pH were required for proteosynthesis. Maximum synthesis took place at pH values ranging from 9 to 11, whereas the pH optimum for amide bond hydrolysis approximates to 7.8.[14] Apart from a few exceptions, the reaction yields were proportional to the nucleophilicity of the amine components and inversely proportional to the solubility of the resulting products. This observation appears intelligible, if one bears in mind, firstly, that nucleophilicity is a measure of the affinity for the acyl-chymotrypsin complex of the nucleophilic acceptor, and secondly that the precipitation of the products both shifts the equilibrium point in favor of the synthesis (*vide supra*, Chapter 5, Section II.A) and, beyond that, prevents the newly formed, insoluble product from undergoing *a posteriori* hydrolysis. Furthermore, the product yields were dependent on the concentration of the amine components. This finding comes as no surprise; if one remembers the strong influence of the educt concentration on the reaction yield (*vide supra*, Chapter 5, Section II.C) and also takes into account the fact that the amine component must compete with water for the deacylation of the acyl-protease complex. In addition, it is a well-known fact that the presence of a nucleophile inhibits the chymotryptic hydrolysis of those substrates which contain the same nucleophile as leaving group.[15] Thus, in the system of Morihara and Oka,[13] *a posteriori* hydrolysis of the newly emerged products was inhibited because the amine component was identical with the potential leaving group. The extent of this so-called secondary hydrolysis increased with decreasing concentration of the starting materials, in particular, of the amine components. However, the initial rate of synthesis was markedly higher than that of the secondary hydrolysis and the highest obtainable product concentration was usually reached within 2 to 5 min; and so this undesired side reaction could be neglected. During these enzymatic syntheses, the authors observed that the secondary specificity of chymotrypsin for the P'_1-position of the amine component closely resembled its leaving group specificity in proteolysis,[16,17] i.e., hydrophobic and bulky amino acid residues were the best choices. Free amino acids and their ester derivatives were found to be poor acceptor nucleophiles, whereas amino acid and peptide amides as well as free peptides and their esters were better suited as amine components. The failure of the free amino acids to serve as adequate nucleophile acceptors seems plausible in view of the rather high basicity of their α-amino groups, the pK_a value of which generally exceed those of the corresponding amides and of comparable di- and tripeptides.[18] In this context, one should bear in mind that, as far as proteosynthesis is concerned, deacylation of an acyl-chymotrypsin intermediate requires nucleophilic attack of an amine component. The effective nucleophile concentration, however, is determined solely by the degree of deprotonation of the α-amino groups at a given pH. In contrast, the low efficiency of the syntheses in which amino acid esters were used in place of the corresponding amides, cannot be explained in terms of different nucleophilicities or solubilities. Even the occurrence of an α-chymotrypsin-mediated esterolysis cannot fully account for the poor yields because peptide esters were found to be useful nucleophilic acceptors. It is, however, conceivable, that an amino acid ester is less compatible with the geometry of the S_1'- or S_2'-site of α-chymotrypsin than the corresponding amides. This possibility is supported by the finding that the synthesis proceeds much more slowly when D-leucine amide is used as amine component instead of L-leucine amide, and that H-Pro-NH$_2$ does not react at all. Again, these observations can hardly be consequences of differential solubilities or nucleophilicities, however, they may also be explained by sterical incompatibilities.

These results clearly demonstrate the importance of the secondary specificity of α-chymotrypsin — that is to say, of the secondary enzyme-substrate interactions — for the enzymatic peptide synthesis.

In an additional study on chymotrypsin-controlled peptide bond formation,[19] Oka and Morihara employed N^α-blocked amino acids and peptides as acyl group donors which had free α-carboxyl moieties as opposed to the esterified acyl donors used during the aforementioned report.[13] A comparative analysis of the data given in both studies provided the following picture; in the presence of esterified acyl donors the reaction rate is very high at a pH of approximately 10. Contrary to that, the reaction proceeds very slowly, despite the presence of a quite high enzyme concentration, at pH 7.0, the optimum for peptide synthesis when N^α-protected amino acids and peptides with free α-carboxyl groups serve as acyl donors. As the pH optimum for peptide synthesis did not significantly differed from that for peptide hydrolysis (pH 7.8) and as a consequence of the prolonged reaction time (20 hr), the newly formed products are always at the risk of undergoing secondary hydrolysis. This problem could only be overcome by adding the amine components in a large molar excess. In view of the above findings, esterified acyl group donors appear to be highly preferable over those having a free α-carboxyl group. In fact this is true not only for chymotryptic syntheses but also for trypsin- and papain-catalyzed peptide syntheses (vide infra, this chapter, Sect. III and VIII, respectively). These results do not appear surprising, if one takes into consideration that the formation of a covalent acyl-enzyme complex by the aforementioned proteases represents an esterification process (vide supra, Chapter 4), that can be reduced to an energetically favored transesterification process in the presence of an a priori esterified acyl-group donor. Conversely, the use of an acyl donor with a free α-carboxyl group would require a thermodynamically less favorable de novo esterification.

It is indeed remarkable that according to the present[19] and the preceding report,[13] α-chymotrypsin is capable of catalyzing peptide bond formation in the presence of D-leucine amide, i.e., the optical antipode to the naturally occurring L-amino acid. Obviously, the stereospecificity of chymotrypsin with respect to the P_1'-site of a given substrate is not as absolute as that observed for the P_1-site.

Similar findings to those described above concerning the efficiency of α-chymotrypsin-catalyzed syntheses with aromatic N^α-blocked, COOH-terminally free amino acids as acyl-donors were reported by Luisi and co-workers.[20,21] During these syntheses which took place at pH 6.5[20] and 7.7[21], respectively, dipeptide amides, amino acid methyl- and ethylester were used as amine components. The low yields in tripeptide amides obtained in some of these reactions[20] were possibly caused by side reactions of the amine components (in the present case of the dipeptide amides) during the prolonged incubation period of 2 to 5 days. Thus, since the dipeptide amides have an aromatic amino acid residue in the C-terminal position, they may be chymotryptically deaminated and/or transformed via ring closure to the respective diketopiperazines.

A further undesired side effect of chymotrypsin, namely its esterase activity, may be held responsible for the poor yields obtained during those syntheses which took place in the presence of amino acid esters such as H-Gly-OMe, H-Gly-OEt, or H-Ala-OEt as nucleophile acceptors.[21] Although the esterase activities of chymotrypsin toward glycine esters can be considered as being rather small,[22] a major amount of free glycine may have been released, during the extended incubation time (2 to 3 days).

The extremely prolonged incubation period certainly represents a serious shortcoming for those syntheses involving C-terminally underivatized carboxyl components. In so far, these reactions are thermodynamically rather then kinetically controlled. Thus, the positive effect on the product yield of the high initial velocity of aminolysis — relative to the initial rate of hydrolysis — would be largely abolished by the negative influence of an enforced tendency to secondary hydrolysis. Under these circumstances, only the low solubility of the products permits significant synthesis to proceed.

In another study dealing with enzymatic fragment condensation, it was shown that peptide syntheses could also be successfully performed via δ-chymotrypsin-catalysis.[23] (δ-Chymotrypsin differs from α-chymotrypsin in that the B- and C-chains are still connected.) Starting with N$^\alpha$-acetylated, α-carboxyl-methylated acyl-group donors and α-carboxyl-amidated amine components the resulting tetra-, penta-, and hexapeptides could be obtained in good yields.

In an additional study on δ-chymotrypsin-assisted peptide bond formation[24] Bizzozero et al. employed Z-Tyr-OH and Z-Phe-OH, respectively, as carboxyl components and amino acid or dipeptide amides as amine components to prepare Z-protected di- and tripeptide amides. The peptidic synthons, the N$^\alpha$-deacylated form of which was to serve as amine component in elastase-catalyzed fragment condensation (see below), could be obtained in high yields as crystalline precipitates. The proteosynthetic activities of chymotrypsin are not solely confined to substrates possessing an aromatic amino acid residue in the P$_1$-position. This was demonstrated by Tominga et al.[25] who prepared in the presence of α-chymotrypsin the protected dipeptide Boc-Asn-Cys(Bzl)-N$_2$H$_2$Ph, a fragment related to oxytocin, from Boc-Asn-OEt and H-Cys(Bzl)-N$_2$H$_2$Ph. Beyond that, Mancheva et al.[26] reported on the incorporation of cysteine sulphonamide (H-Cys(O$_2$NH$_2$)-OH) via α-chymotryptic catalysis into several oligopeptides. During these studies the cysteine sulphonamide residues were incorporated both into the P$_1$- and the P$_1'$-position of the resulting peptides which displayed antibacterial activities.

To counter the rather low solubility of several donor esters in aqueous solutions, Morihara and Oka[13] performed the α-chymotrypsin-mediated coupling reactions in dimethyl formamide-water mixtures. By adding organic co-solvents the authors certainly induced a favorable shift of the ionic equilibria (*vide supra*, Chapter 5). However, the dark side of the picture, the reduction in the catalytic activity of chymotrypsin as caused by the presence of these solvents, had to be compensated by increasing the concentration of the enzyme. A promising approach to obviate the deleterious effects of nonaqueous co-solvents on protease activity[27,28] was suggested by Kuhl et al.[29] and by Semenov et al.[30] who allowed enzymatic peptide synthesis to take place in biphasic organic-aqueous systems. The aqueous layer of this system offers ideal conditions for the protease to fully exert its catalytic activities, which are then hardly impaired by the second phase consisting of water-immiscible organic solvents.

α-Chymotryptic peptide syntheses in biphasic systems have been performed successfully by Martinek and co-workers[30-32] and, in particular, by Jakubke and collaborators.[29,33-37]

The relative amount of water in these systems can be largely reduced to the advantage of peptide synthesis.[31] Thus, Kuhl et al. were able to demonstrate the efficient synthesis via chymotrypsin catalysis of the tripeptide Ac-Leu-Phe-Leu-NH$_2$ in 92% yield in a biphasic system composed of carbonate buffer, pH 10, 98% (v/v) trichloroethylene, and only 2% (v/

v) water .[33]   Although no condensed product was obtained in the complete absence of water, it is indeed remarkable that the mere addition of the decahydrate of sodium carbonate enables the reaction to proceed yielding the target peptide in 91% yield.

An interesting variant of peptide synthesis in biphasic systems was recently presented by Lüthi and Luisi.[38] In this case, the aqueous phase was embedded in the interior of reverse micelles, which in turn were surrounded by the organic phase. (The reverse micelles were formed by an anionic surfactant.) Using this novel biphasic system with the chymotrypsin catalyst dissolved in the polar core of the micelles, the authors succeeded in synthesizing a dipeptide and a tripeptide in good yields. (For detailed reports on peptide syntheses in biphasic systems cf. References 39 and 40).

An alternative way to counter the rather low solubility in aqueous solutions of hydrophobic N^{α}-protected acyl group donor esters was proposed by Kuhl et al.[41,42] The authors employed hydrophilic ''solubilizing'' ester moieties in place of the commonly used ethyl or methyl esters. Several water-soluble N^{α}-acylated peptide 4-sulfobenzyl esters[41] and 2-thiosulfato-ethyl esters[42] were successfully used as carboxyl components for the chymotrypsin-catalyzed synthesis of N^{α}-protected peptide amides.

While all the reactions discussed so far were catalyzed by free ''mobile'' proteases, Könnecke et al.[34] initiated the use of ''immobilized'' α-chymotrypsin for peptide synthetic purposes. This technique was remarkable for its economy since the enzyme could be used repeatedly while still enabling successful syntheses. Könnecke et al.[36] further reported on peptide syntheses in biphasic systems during which ''freely mobile'' chymotrypsin and Z-Leu-Phe-OMe served, respectively, as catalyst and acyl group donor while immobilized leucine amide was used as amine component. However, peptidic synthons could not be formed unless the spacers, through which the amine component was covalently attached to an insoluble silica support, had been considerably lengthened. In spite of that, the yield was rather moderate, presumably because the amine component — due to its restricted mobility — was out competed by the water component of the reaction mixture for the deacylation of the Z-Leu-Phe-chymotrypsin complex. The resulting product of hydrolysis, Z-Leu-Phe-OH, was per se a potential acyl group donor. However, it possessed a free α-carboxyl group, and as therefore might be expected (*vide supra*), it did not represent an efficient acyl group donor in chymotrypsin-controlled syntheses. Obviously, the procedure described by Könnecke et al.[36] bears some analogies to solid-phase peptide synthesis as initiated by Merrifield.[43] The authors thereby presented a practicable means of fragment condensation on solid supports without running the risk of racemization. This comes true in spite of the mediocre outcome of the chymotryptic synthesis; as will be shown below (this chapter, Section X), the product yield could be considerably increased by using the metalloprotease thermolysin in place of chymotrypsin.

Another technique for the preparation of peptides via protease catalysis was developed by Könnecke et al.[44] after the model of liquid-phase peptide synthesis.[45] Amino acids and dipeptides esterified to polyoxyethylene — a soluble polymeric support usually used in liquid-phase peptide syntheses — served as substrates during chymotryptic peptide bond formation. The coupling yields were rather moderate when N^{α}-protected amino acid or dipeptide polyoxyethylene esters were used as acyl group donors, while good yields were obtained with the amine components esterified to the soluble polymeric support. The novel approach to enzymatic peptide synthesis mainly profits from the solubilizing effects of the polyoxyethylene supports. As a result, the solubility of hydrophobic substrates in aqueous reaction media is increased without resorting to the solubilizing power of organic co-solvents (see above).

III. TRYPSIN

Besides chymotrypsin, trypsin is another prominent member of the family of functionally and structurally related serine proteases. Homologies in the primary structure and molecular architecture of chymotrypsin and trypsin[46-48] are probably responsible for the analogies in their catalytic mechanisms. The principal difference relates to their primary specificities. The action of trypsin is strictly confined to peptide linkages, the carboxyl group of which is contributed by a basic amino acid such as lysine or arginine.[49] The striking similarities between both of these serine proteases prompted Oka and Morihare[50] to explore whether trypsin might, like chymotrypsin, represent a suitable catalyst in the field of synthetic peptide chemistry. Indeed, the use of trypsin enabled the authors to prepare a series of model peptides. In these studies N^α-protected amino acid or -peptide esters, which could contribute the acyl portion of the future peptide bond with an arginine or a lysine residue, were used as acyl group donors and various amino acid and peptide derivatives served as nucleophilic acceptors. The conditions prevailing during the enzymatic reactions were similar to those described for chymotrypsin-controlled couplings.[13] To achieve optimal results, considerably more enzyme and a more alkaline pH (10 to 10.5) relative to tryptic hydrolysis were required. Hydrophobic and bulky amino acid amides were found to be the most appropriate nucleophilic acceptors, and amino acid amides or free tripeptides turned out to be better nucleophiles than free dipeptides. Neither free leucine nor the valine *t*-butyl ester were recognized as appropriate amine components indicating that free or esterified amino acids are less suitable for synthesis. By analogy with chymotrypsin (*vide supra*), trypsin does not possess a strict stereospecificity, for the P'_1-position (cf. Figure 1) of the amine component, i.e., D-Leu-NH$_2$, could be used as nucleophilic acceptor, although the L-enantiomeric form of leucine was by far the better amine component. The reaction yields improved with decreasing solubility of the products, and with increasing concentrations of the amine components. The trypsin-controlled synthesis of Bz-Arg-Leu-NH$_2$ was completed within 5 min, but *a posteriori* hydrolysis commenced once the reaction period had exceeded 1 hr. By comparison, secondary hydrolysis was negligible during the α-chymotrypsin-mediated synthesis of Ac-Phe-Leu-NH$_2$, even though the reaction was allowed to continue for 20 hr. These results seem plausible if one considers the differing solubilities in the reaction medium of Bz-Arg-Leu-NH$_2$ (ca. 40 mM) and of Ac-Phe-Leu-NH$_2$ (9.2 mM). In general, products of trypsin-controlled reactions will be more prone to *a posteriori* hydrolysis than those generated by α-chymotrypsin-catalysis, because the solubility in aqueous solutions of basic amino acids significantly exceeds that of aromatic amino acid residues. Consequently, during tryptic syntheses a high amount of the desired peptide must be formed within a short period so that the reaction can then be stopped in order to prevent secondary hydrolysis. This means that, as a rule, trypsin-catalyzed peptide syntheses proceed under kinetic control (cf. Chapter 5, Section III).

   With respect to N^α-protected acyl components possessing free instead of esterified α-carboxyl-groups, trypsin displayed a behavior comparable to that of chymotrypsin.[51,52] During these studies leucine-[51] and valine- derivatives[52] were used as amine components and the optimal pH value for peptide bond formation was found to be 6.5 to 7.0. Reasonable yields could not be obtained unless both the concentration of the amine components was considerably increased relative to that of the acyl components and the incubation period was significantly extended. The following example illustrates the differential efficiency of free and esterified carboxyl components in peptide synthesis. Thus, a 40-fold excess of H-Leu-NH$_2$ over Z-Arg-OH was required to prepare Z-Arg-NH$_2$ at pH 6.5 in 49% yield after 20 hr whereas at pH 10.4 equimolar concentrations of Bz-Arg-OEt and H-Leu-NH$_2$ in the presence of half the concentration of trypsin was sufficient to obtain the very same dipeptide in 72% yield within 2 min.[50] The comparatively low coupling efficiency of the former reaction was presumably brought about by secondary hydrolysis which, as mentioned earlier, started after an incubation period of only 60 min.[50] An additional explanation may be as follows; the actual ''operational'' nucleophile concentration is represented by the deprotonated fraction of the total leucine amide (pK_a = 8.70 [18]) concentration, the percentage of which is certainly far larger at a pH value of 10.4 than at a pH value of 6.5 to 7.0. High concentrations of organic cosolvents promoted the synthesis due to their ability to favorably shift the pK_a value of the ionogenic reagents. However, the fraction of nonaqueous co-solvent could not be increased *ad libitum*.[52] Beyond a critical concentration any further addition of organic solvents led to a decline of the reaction yields which was obviously caused by the denaturing effect of the nonaqueous media. The coupling efficiency of the tryptic syntheses also depended on the chemical nature of the leucine derivatives serving as nucleophiles. Thus, the degree of synthesis was found to obey the following order: leucine anilides > leucine esters > leucine amides. That is, the reaction yields increased with decreasing basicity of the amine components. This point may be neglected when a reaction takes place at pH 10.4 where the amino groups are largely deprotonated; but it may play a decisive role in the pH range of 6.5 to 7.0, when a significant portion of the "amine component" exists in a protonated form, which cannot participate in the peptide bond forming step. It is noteworthy, that H-Leu-OEt and H-Leu-OBu^t are more suitable nucleophilic acceptors for trypsin-catalyzed peptide synthesis than H-Leu-NH$_2$, whereas the opposite proved to be the case for chymotrypsin-mediated reactions.[19] These findings may be attributed to differing secondary specificities and/or esterase specificities exerted by the two proteases; that is to say, thus, whereas leucine esters are hardly affected by trypsin, they are readily attacked by the esterase activities of chymotrypsin.

As observed by Widmer et al.[53] and Mancheva et al.,[26] the "proteosynthetic" primary specificity of trypsin, which is associated with the P_1-site of the tryptic substrates, is not confined to basic amino acid residues. The product yields of trypsin-catalyzed peptide syntheses reported by the authors indicated that alanine-, serine-, histidine-,[53] as well as cysteine sulphonamide — Cys(O$_2$NH$_2$) — residues[26] in the P_1-position were recognized as tryptic substrates by the protease.

Peptide bond formation by means of immobilized trypsin was studied by Könnecke et al.[54] The authors were able to reutilize the protease, which was covalently attached to insoluble carriers, as many as five times. The model peptides prepared by this technique were invariably obtained in good yields.

IV. SUBTILISIN BPN'

An additional serine protease, the proteosynthetic potential of which has been explored, is subtilisin BPN', a protease which is isolated from *Bacillus subtilis*. In contrast to chy-motrypsin and trypsin, this protease is known to display a broad primary specificity,[55] though it demonstrates a chymotrypsin-like preference for aromatic amino acid residues in the P_1-site of its substrates. Despite the significant structural differences between chymotrypsin and subtilisin BPN'[56] the two proteases closely resemble each other in their catalytic mechanisms.[57]

By synthesizing several model tetrapeptides, Isowa et al. investigated the potential of peptide bond formation via subtilisin BPN'-catalysis.[58] Dipeptides of the general form Z-Phe-X-OH served as carboxyl components and the dipeptide H-Phe-Val-OBu^t was used as the amine component throughout these studies. Only moderate yields were obtained at pH 7.2 when the X-position was occupied by valine-, tyrosine-, or N^G-nitro-arginine moieties. In general, the efficiencies of these syntheses were significantly lower than those observed during comparable thermolysin-catalyzed reactions (cf. this chapter, Section X). With gly-cine- and serine residues located in the above X-site, condensed products were not produced at all.

On the other hand, Isowa et al. in a further study[59] on subtilisin BPN'-catalyzed peptide synthesis showed that the protease could be used to good advantage for the coupling of two conventionally prepared angiotensin fragments, namely, the dipeptide Boc-Val-Tyr(Bzl)-OH and the tetrapeptide H-Val-His(Bz)-Pro-Phe-OEt. However, the resulting hexapeptide, which was obtained in a good yield, lacked the C-terminal ethyl ester; a side reaction, which was obviously caused by an esterase activity elicited by subtilisin BPN′.

In a comparative study Morihara and Oka[60] investigated the influence of the chemical composition of various carboxyl- and amine components on subtilisin-catalyzed syntheses. The efficiency of the acyl group donors, which were N^α-protected, but α-carboxylate unprotected, increased with growing chain length. Amino acid anilides were the best choice as acceptor nucleophiles whereas amino acid amides and esters were inadequate, even if their concentration exceeded that of the respective carboxyl component by more than ten times. These findings may be explained both by the susceptibility of the esterified amine component to the esterase activities of the proteases and by the decreased basicity and solubility of the anilides relative to the amides.

N^α-protected di- and tripeptide methyl esters and amino acid *p*-nitroanilides were used by Voyushina et al.[61] as carboxyl and amine components, respectively, to prepare via subtilisin-catalysis a series of tri- and tetrapeptides. When Z-Ala-Ala-OMe and Z-Ala-Ala-Leu-OMe served as carboxyl components, the authors observed that the coupling efficiency was strongly influenced by the chemical nature of the amine components in the following order: H-Ala-pNA > H-Val-pNA > H-Leu-pNA > H-Phe-pNA. The product yields were signficantly reduced, if the above esterified acyl group donors were replaced by those having a free α-carboxyl group. So far, subtilisin elicits a behavior similar to that of chymotrypsin and trypsin (see above).

V. ELASTASE

Although the serine protease elastase does not exhibit a stringent primary specificity, it acts preferentially upon peptidic substrates the P_1-position of which is occupied by alanine, serine, methionine, or phenylalanine residues.[62] Furthermore, kinetic studies indicate that secondary elastase-substrate interactions significantly influence the catalytic activity of the proteases.[63,64] The "proteosynthetic" potential of elastase was studied by Voyushina et al.[61] The authors used Z-Ala-OMe, Z-Ala-Ala-OMe, and Z-Gly-Ala-OMe as carboxyl components while H-Leu-pNA and H-Ala-pNA served as amine components. The effect of introducing an additional alanine or glycine residue into the P_2-position of the carboxyl component on elastase-catalyzed peptide bond formation was characterized by considerable increases in the product yields. The importance of secondary protease-substrate interactions is illustrated by the following example: with Z-Ala-OMe and H-Leu-pNA the expected dipeptide could be obtained via elastase-catalysis in modest yield (49%); however, when Z-Ala-Ala-pNA was used in place of Z-Ala-pNA elastase-mediated peptide bond formation gave a high yield of the desired tripeptide (73%). The prominent role played by the secondary specificity of elastase in peptide–bond-forming steps is also emphasized by a report of Bizzozero et al.[24] These studies involved the coupling of the tripeptide N^α-Ac-Ala-Ala-Ala-OEt with various di- and tripeptide amides in the presence of porcine elastase. Among the dipeptide amides serving as amine components H-Tyr-Ala-NH$_2$ was effectively coupled with the above tripeptide ester to give the targeted pentapeptide in 68% yield, while H-Tyr-Gly-NH$_2$ turned out to be a rather poor choice; i.e., the corresponding pentapeptide could only be obtained in 12% yield. The striking difference in coupling efficiency suggests specific secondary interactions between the active site of the protease and the amino acid residue in the P_2'-position of the amine component. Contrary to that, the P_3'-site does not appear to be of major importance in elastase-catalyzed peptide bond formation; that is to say, the replacement of the dipeptide H-Tyr-Ala-NH$_2$ by the tripeptide H-Tyr-Ala-Ala-NH$_2$ had relatively little effect on the coupling yield.

VI. THERMITASE

Thermitase, another serine protease which was isolated from *Thermoactinomyces vulgaris*, was introduced by Könnecke and Jakubke[65] as a potentially useful biocatalyst in peptide synthetic chemistry. Thermitase-catalyzed syntheses were performed in biphasic systems using dipeptides of the general form Z-X-Y-OMe as carboxyl components and leucine amide as amine components. In the case where the X-position of the substrate was occupied by a leucine residue, the reaction yields decreased according to the following order with Y = Met > Tyr, Leu > Phe > Ala. When substrates of the series Z-X-Phe(Ala)OMe were used the preference of thermitase for the X-position decreased as follows: Pro, Val > Ala > Leu > Gly. The protease obviously displays a primary specificity which is directed toward the P_1-site and a distinct secondary specificity for the P_2-site of its substrates.

VII. CARBOXYPEPTIDASE Y (CPD-Y)

In contrast to the serine proteases mentioned so far, carboxypeptidase Y is an exopeptidase that commonly catalyzes the sequential removal of amino acids from the C-terminal end of a protein- or peptide chain. This protease, which can be obtained from Baker's yeast, exhibits a broad primary specificity[66] and was suggested as a catalyst for peptide synthetic purposes by Widmer and Johansen.[67] The authors used Bz-Ala-OMe and Z-Phe-OMe as acyl donors and a multitude of free and derivatized amino acids as amine components. The syntheses were conducted at pH 9.6 or 9.7, because CPD-Y displays strong esterase activity at pH values above 9, for this reason the carboxyl components had been methylated, whereas its peptidase activities are greatly reduced whithin this pH range.[68] The reaction yields depended crucially on the chemical nature and the concentration of the amine components which by far exceeded the concentration of the respective carboxyl components. Amino acid amides and -hydrazides were found to be more suitable for synthesis than the corresponding free amino acids. Furthermore, the latter were more efficiently incorporated when carrying a hydrophobic or a positively charged side chain. In contrast, amino acid residues with carboxyl-, hydroxyl-, and amide side-chain functions proved to be poor amine components. Clearly, these three functionalities prevented a favorable interaction of the respective amine component with the S_1'-position of the active site of the protease.

Structural patterns of the amine components also played a decisive role in the following studies.[69-71] Again it was observed that amino acid amides are better nucleophilic acceptors than free amino acids. Additionally, D-amino acids, dipeptides, and secondary amides were shown to be inappropriate amine components.[69] The use of amino acid alkyl esters as amine components in the presence of N^α-blocked amino acid and peptide methyl esters as carboxyl components permitted the synthesis of the intended peptides. However, in many cases by-products with an extended chain length were also created.[70] In fact surplus amino acid residues were incorporated into the products by random oligomerization of the amine components; the methyl esters of hydrophobic amino acid residues in particular being involved in this side reaction, whereas hydrophilic amino acid methyl esters were oligomerized only to a minor extent. The degree of oligomerization, however, could be reduced both by enlarging the alkyl ester moiety of the amine component and by employing phenylmercuric chloride-treated carboxypeptidase Y.[71] In addition, this latter study revealed that benzylated ester donors may serve the function of a carboxyl component more efficiently than the corresponding methyl esters. This effect was observed for instance with the benzylated alanine derivative Bz-Ala-Bzl which proved to be strikingly superior as an acyl group donor in comparison to its methylated analog Bz-Ala-OMe.

In another study, Breddam et al. described transpeptidation phenomena they had observed during CPD-Y-catalyzed syntheses.[72] Using benzoyldipeptide amides as carboxyl components and valine amide as amine component, the authors showed that the carboxyl component was either enzymatically deaminated followed by valine amide incorporation in place of the released amide group, i.e., the desired peptide-chain-lengthening step took place, or the peptide bond of the carboxyl component was enzymatically cleaved and the released amino acid amide subsequently replaced by valine amide, i.e., an undesired transpeptidation step preventing the synthesis of the targeted tripeptides occurred. Obviously, these randomly occurring proteolytic activities of CPD-Y jeopardize a stepwise peptide synthesis — starting from the α-amino terminus — that, in the presence of an exopeptidase, could formally be a promising prospect for enzymatic syntheses.

Studies on CPD-Y-mediated peptide syntheses in biphasic aqueous-organic systems were performed by Kuhl et al.[73] In most cases N^α-protected di- and tripeptide amides could be prepared in excellent yields. N^α-Boc- and -Z-blocked amino acid- and dipeptide methyl esters were employed as carboxyl components while valine- and leucine amides provided the amine components.

VIII. PAPAIN

Papain was the very first protease whose proteosynthetic capacity was exploited for the preparation of a well-defined peptide.[74] The catalytic mechanism of papain, a cysteine protease isolated from *Carica papaya,* involves the formation of a covalent acyl-enzyme complex. However, in contrast to the previously mentioned serine proteases, the acyl function of the substrate is bound to an active-site cysteine residue via a thioester linkage (*vide supra,* Chapter 4). Papain displays a rather broad substrate specificity, but nevertheless, it exhibits a preference for bulky hydrophobic amino acid residues in the P_2-position of a given substrate.[62] Given that its bias in favor of the P_2-site is a feature of its primary specificity, then the secondary specificity of papain is associated with the P_1-position, where bulky, hydrophobic amino acid residues are favored, if only to a minor extent.[75,76] Thus, the P_1-site, toward which the primary specificity of many proteases is normally expressed, can be occupied by a variety of amino acids.[77] Due to the structural stability of papain over a wide pH range it does not come as a surprise that papain-controlled peptide syntheses have been performed within a broad pH spectrum which ranges from 4.7[78] to 9.5.[79] The pH optimum for peptide bond formation under otherwise comparable conditions obviously depends on the chemical nature of the substrates. For instance, Fox and co-workers[80,81] observed that N^α-benzoylated leucine, alanine, valine, and glycine residues were most efficiently coupled with aniline near pH 5.0. However, optimal yields for the condensation of the partially protected Bz-Tyr(Bzl)-OH, and aniline were obtained near pH 6.2, while the amount of the condensed product obtained at pH 5.0 was negligible.

In more recent studies on papain-controlled peptide bond formation Isowa et al.[58] described the synthesis, at pH 6.6, of model peptides starting from N^α-Z-protected amino acids or peptides and amino acid- or peptide *t*-butyl and diphenyl methyl esters. As far as the specificity of papain for the P_1-site was concerned, hydrophilic or small hydrophobic amino acids were generally found to be more appropriate as acyl group donors than bulky hydrophobic amino acid residues. Significantly improved reaction yields resulted however if the latter occupied the P_2-site of the substrate. In general, the strongly hydrophobic amino acid and dipeptide esters used during this study proved to be very suitable nucleophilic acceptors. Nevertheless, some remarkable exceptions were observed. Thus although the amine components H-Phe-Phe-OBu^t and H-Phe-Val-OBu^t successfully reacted with several acyl group donors, they did not give any condensed product in the presence of the carboxyl components Z-Phe-OH and Z-Phe-Tyr-OH, respectively. On the other hand, Z-Phe-OH and Z-Phe-Tyr-OH represented adequate acyl group donors themselves, since they could be very efficiently coupled with H-Val-ODMP and H-Phe-ODMP, respectively. These failures can neither be explained by a more or less pronounced solubility of the products nor by the chemical nature of the respective carboxyl- or the amine components which individually were obviously compatible with the corresponding S-positions at the active site of the enzyme. Individual compatibility is certainly a prerequisite for successful binding, however, it does not appear to be sufficient; productive binding presumably also requires a mutual compatibility of the acyl group donor and the acyl group acceptor and the inability of papain to catalyze the desired reaction may be intelligible in terms of negative cooperativity. Thus the binding of the first substrate might induce a conformational change in the papain molecule thereby effecting the binding of the second substrate.

The outcome of a systematic study on the enzymatic coupling of Z-Arg-OH and H-Leu-X (X = NHC_6H_5, OBu^t, OEt, NH_2) revealed that, under the conditions used, papain exhibited proteosynthetic activities markedly superior to those of trypsin,[51] notwithstanding the commonly known preference of trypsin for arginine residues in the P_1-site of its substrates. The optimal pH value for papain-catalyzed peptide bond formation was found to be 5.5 and reactions were driven toward synthesis by high concentrations of the amine components and by the presence of organic co-solvents. The efficiency of the leucine derivatives followed the order: X = NHC_6H_5 > OBu^t, OEt > NH_2. These findings appear plausible, because this series roughly reflects both the decreasing nucleophilicity of the respective amine components and the increasing solubility of the resulting Cbz-Arg-Leu-X peptides. As far as esters, in particular ethyl esters, or amides, were concerned, the esterase or amidase activity of papain may contribute to a reduction of the product yields.

In another work Morihara and Oka[60] studied comparatively the influence of the chemical nature of various carboxyl- and amine components on the coupling yields of papain-mediated reactions. At pH 5.0, where the dipeptide Z-Ala-Leu-ODPM could be prepared in optimal yields from Z-Ala-OH and H-Leu-ODPM, the degree of synthesis using Z-Ala-OH or Z-Phe-OH as carboxyl components and H-Leu-NH_2 or H-Leu-OEt as amine components was negligible. In contrast high coupling yields were obtained when the amide or the ethyl ester of leucine were replaced by the anilide or the diphenylmethyl ester. Despite the good yields obtained during the coupling of Z-Ala-OH or Z-Phe-OH and H-Leu-NHC_6H_5, the usefulness of leucine anilide as amino component is somehow impaired because the syntheses were accompanied by the formation of oligomeric by-products. The latter were generated through a concerted action of the hydrolytic and proteosynthetic properties of papain, i.e., the anilide group of the amine component was enzymatically cleaved and the resulting free leucine molecules were additionally incorporated in the growing peptide chain via papain catalysis. Moreover, Z-Gly-OH, Z-Glu-OH, and Z-Val-OH were also found to be useful acyl group donors when reacted with H-Leu-ODPM, whereas Z-Pro-OH, Z-D-Ala-OH, Z-Gly-Phe-OH, or Z-Gly-Pro-Leu-OH failed to give condensed products under comparable conditions. Nevertheless, the above di- and tripeptides were suitable carboxyl components, when they were allowed to react with H-Leu-NHC_6H_5. Although, the formation of oligomeric by-products was again observed, the results of the foregoing study are indicative of the synthesis-promoting effect of the anilide moiety. Unfortunately, this group is not selectively removable, and it is therefore not advisable to use it as protecting group during the assembly of larger peptides. However, the phenylhydrazide protector group, which is chemically related to the anilide group, not only promotes protease-catalyzed peptide bond synthesis (*vide infra*, Chapter 8, Section I) but it also meets the requirements of an appropriate blocking group; i.e., it can be selectively removed.

The widespread ''specificity'' of papain is further illustrated by a study dealing with the enzymatic incorporation of the noncodogenous, rare amino acid γ-carboxyglutamic acid (Gla) into synthetic peptides.[82] Starting from heterochiral Z-DL-Gla-OH serving as carboxyl component and several amino acid phenylhydrazides serving as amine components, Čeřovský and Jošt enantioselectively prepared via papain-catalysis a series of homochiral dipeptides of the form Z-L-Gla-X-N$_2$H$_2$Ph, where X denotes an unspecified amino acid residue. The product yields were crucially dependent on the chemical nature of the respective amine component; the coupling efficiency of the amino acid occupying the X-position followed the order: Leu > Phe > Met > Ala > Val > Asn.

The proteosynthetic capacities of papain in the coupling of peptide fragments were demonstrated during the preparation of an angiotensin octapeptide[59] and of several enkephalin derivatives.[83] In most of the cases, the coupling of the peptide subunits, which were synthesized prior to use by chemical procedures, resulted in very good product yields.

Heretofore, all the studies on papain-catalyzed peptide syntheses were based on acyl group donors the α-carboxyl groups of which had not been substituted before use. However, Döring et al. observed that, in analogy to the chymotrypsin-mediated syntheses discussed above, papain-catalyzed syntheses could also proceed more efficiently when esterified acyl group donors were used.[84] Whereas these syntheses were performed at pH 5.5, that is, in a pH range which also favors the peptidase activities of the enzyme, Mitin et al. reported on successful peptide syntheses via papain-catalysis at pH values ranging from 8.0 to 9.5.[79] In the latter case, the authors used N$^{\alpha}$-blocked amino acid methyl esters as carboxyl components thereby exploiting in elegant fashion the pH-dependent functional features of papain, which shows largely reduced peptidase activities at pH values greater than 8 while retaining considerable esterase activity.[85]

An attempt to use esterified acyl group donors in the pH range from 7.3 to 8.9 for peptide synthetic purposes in the presence of immobilized papain was less successful.[54] Obviously, the cleavage of the ester bond of the carboxyl component did not proceed appreciably faster than the secondary hydrolysis of the newly formed peptide bonds. Consequently, the product yields were rather unsatisfactory.

A comparison between papain-catalyzed peptide synthesis in biphasic aqueous-organic systems and in monophasic aqueous-organic mixtures was described by Jakubke and collaborators.[29,37,84] The authors demonstrated that enzymatic synthesis in water — water-immiscible organic solvent systems — resulted in a markedly higher coupling efficiency than in monophasic aqueous-methanolic mixtures. In this context, tetrachloromethane and ethyl acetate appeared to be the most promising organic solvents. As far as the percentage of the organic solvent was concerned, the highest coupling yields were achieved when the tetrachloromethane fraction came to approximately 40% of the total volume of the biphasic system.

In summary, it can be concluded from the results of papain-catalyzed syntheses of model peptides, that the protease accepts a broad structural range of substrates and therefore mediates the formation of a variety of different peptide bonds. Nevertheless a caveat should be borne in mind: the relative lack of specificity of papain raises the possibility that pre-existing peptide bonds in a substrate may become subject to secondary hydrolysis.

IX. PEPSIN

Pepsin, a protease found in the gastric juice, is the best-known member of the family of the so-called "acid proteases". Since the β-carboxylate functions of two aspartic acid residues are involved in its catalytic mechanism,[86] pepsin, which exhibits a pH optimum ranging from 2 to 3, is also designated as an aspartate protease. Although it is considered a protease of broad structural specificity, pepsin elicits a striking preference for those substrates whose P$_1$- and P$'_1$-positions are occupied by a phenylalanine- and an aromatic amino acid residue, respectively.[87] There still exist some uncertainties about the molecular mechanisms of the peptic catalysis. In particular, the question of whether or not an intermediate acyl-enzyme- and/or an imino-enzyme complex is formed during the catalytic process has not yet been conclusively resolved.[86]

The proteosynthetic capacities of pepsin had already been recognized in connection with the early studies on the plastein reactions.[88] However, a detailed description of the molecular processes governing the pepsin-controlled plastein reactions and a characterization of the resulting products were only concluded some decades later when in the 1960s Wieland and collaborators extensively explored these topics[89] (*vide infra,* Chapter 9).

More recently Isowa et al.[58] reported on the pepsin-assisted syntheses of several tripeptides. Although the authors could couple the dipeptides Z-Leu-Phe-OH and Z-Phe-Tyr-OH with H-Phe-ODPM in good yields, they failed in their attempt to condense Z-Val-Tyr-OH and the above amine component, H-Phe-ODPM. This finding demonstrated that the success of a peptic synthesis is not only dependent upon the P_1- and $P_1{}'$-site of a given substrate, the secondary specificity, as associated with the P_2-position, also plays a decisive role.

In another study, Morihara and Oka[60] elucidated the effect of various leucine-containing amine components on the coupling efficiency of peptic proteosynthesis at pH 4.5. Using Z-Phe-OH or Z-Gly-Phe-OH as acyl donors H-Leu-NH_2 and H-Leu-OEt proved to be poor choices, whereas H-Leu-OBu^t was a moderate, and H-Leu-NHPh a good nucleophilic acceptor. Because pepsin showed neither esterase nor amidase activity, the authors ascribed the poor efficiency of H-Leu-NH_2 and H-Leu-OEt to the comparatively high solubility of the expected products. As already noted for syntheses catalyzed by proteases other than pepsin, amino acid anilides were also able to promote peptide bond formation in the presence of pepsin as mediator.

The pepsin-controlled condensation of N^α-protected aromatic amino acids, or peptides and aromatic amino acid amides or esters at pH 4 was described by Pellegrini and Luisi.[90,91] When H-Phe-NH_2 served as amine component, the authors found that the coupling efficiency was considerably influenced by the nature of the acyl group donors in the following order: Z-Phe-Oh > Z-Trp-OH > Z-Tyr-OH. These results were in good agreement with kinetic data on the hydrolytic specificity of pepsin.[87] Furthermore, it could be shown that H-Phe-NH_2 is a more efficient nucleophilic acceptor than H-Phe-OMe and, that Z-Phe-OH is a better acyl group donor than Ac-Phe-OH, while Boc-Phe-OH was found to be completely unsuitable. The selective behavior of pepsin toward the N^α-protecting groups again emphasizes the important role played by the P_2-subsite of the substrates.

During peptic syntheses of model peptides, comparable to those discussed above, Tsen et al.[92] observed the formation of several by-products. The authors incubated Z-Phe-OH and H-Phe-OBzl at pH 4 in the presence of pepsin and obtained, besides the targeted dipeptide Z-Phe-Phe-OBz1, the tripeptide Z-Phe-Phe-Phe-OBzl. The use of Z-Phe-Leu-OH as carboxyl component and of H-Phe-OBzl as amine component gave similar results in that the tetra-peptide Z-Phe-Leu-Phe-Phe-OBzl was synthesized concurrently with the desired tripeptide. Most probably, an esterase-like activity of pepsin gave rise to the appearance of the "elongated" peptides; i.e., the free amino acid released through the cleavage of the benzylester bond of the amine component was additionally inserted into the growing peptide chain.

Pepsin also acted as catalyst for the synthesis of several enkephalin-peptides[83] as well as for the preparation of some peptide-fragment related to eledoisin and substances P.[93] These enzymatic reactions represented the condensation of peptide subunit which had been prepared via chemosynthesis prior to enzymosynthesis. Last, but not least, Kuhl et al.[94] demonstrated that pepsin-assisted peptide formation can also efficiently take place in aqueous-organic biphasic systems.

X. THERMOLYSIN

Thermolysin, a neutral metalloprotease isolated from *Bacillus thermoproteolyticus* has frequently been used in enzymatic peptide synthesis. The enzyme binds one zinc ion essential for catalytic activity,[95] and four calcium ions, which are required for optimal thermostability.[96] Thermolysin exhibits a strong preference for peptide bonds the imino group of which is contributed by bulky hydrophobic amino acid residues.[55] In addition, the enzymatic activity is further enhanced when the carbonyl portion of the sensitive bond is donated by a hydrophobic residue.[97] Thus, thermolysin represents a rare case, where the primary specificity of a protease is predominantly determined by structural features of the P_1'-site of the substrates. Current information on the mechanism of thermolysin-action is rather contradictory. A general base catalysis has been postulated.[98] Alternatively, catalytic pathways involving both an acyl-enzyme complex — via an anhydride linkage — and an amino-acyl complex are favored by other authors.[99]

The first systematic study on the proteosynthetic potential of thermolysin was published in 1977 by Isowa et al.[58] The protease was shown to catalyze successfully the coupling of different Z-amino acids with H-Phe-Val-OBu^t at pH 7.5 in the following order of efficiency: Z-Gly-OH > Z-Ser-OH > Z-Tyr-OH > Z-Arg(NO$_2$)-OH > Z-Val-OH. Using H-Phe-Phe-OBu^t as a nucleophilic acceptor, Z-Asn-OH, Z-Gln-OH, Z-Leu-OH, Z-Met-OH, Z-Gly-OH, and Z-Tyr-OH were found to be the best choices as acyl group donors in the order indicated. Z-Phe-OH and Z-Thr-OH, however, gave only moderate yields, and Z-Ser-OH did not yield any condensed product at all. As compared to H-Phe-Val-OBu^t, H-Phe-Phe-OBu^t was less effective as an amine component; all the Z-amino acids which reacted with both of these dipeptide nucleophiles gave smaller yields in the presence of H-Phe-Phe-OBu^t. Obviously, there exists a secondary specificity of thermolysin associated with the P_2'-subsite, because substrates having a valine residue in this position are preferred over those having a phenylalanine residue. Thermolysin-controlled condensation of Z-Phe-X-OH and H-Phe-Val-OBu^t revealed that Z-blocked dipeptides were more efficient acyl group donors than the corresponding amino acids derivatives of the form Z-X-OH, when the X-position was occupied by a valine or a tyrosine residue. If, however, X was represented by glycine-, serine, or N^G-nitro-artinine residues, the dipeptides gave lower yields than the corresponding amino acid residues. This means that the presence of an additional phenylalanine unit in the P_2-site affects the reaction yield favorably in some instances and detrimentally in others. Consequently, the thermolysin catalysis is influenced not only by direct enzyme-substrate interactions but also by intramolecular interactions (most probably mediated by the protease) between the amino acid residues located in the P_1- and P_2-positions of the carboxyl components.

In another report Isowa and Ichikawa[100] described the thermolysin-catalyzed synthesis of a series of N$^\alpha$-Z-dipeptide amides and methyl esters at pH 8.0. Isoleucine-, phenylalanine-, valine-, and leucine derivatives could be used successfully as amine components, whereas methionine proved to be suitable only in the amide form. In general, amino acid amides could be used more efficiently as nucleophilic acceptors than methyl esters. However, tyrosine and tryptophan residues whether amidated or methylated were poor acyl group acceptors. The latter finding comes hardly as a surprise because it is known that peptide bonds the imino group of which is donated by tryptophan or tyrosine residues are hydrolyzed at rather slow rates.[97] A variety of N$^\alpha$-Z- or -Z(OMe)-amino acids comprising hydrophobic as well as side chain protected hydrophilic residues functioned effectively as carboxyl components. Even Z-Pro-OH gave a moderate yield when reacted with H-Leu-NH$_2$. However, in spite of their hydrophobic character, valine-, isoleucine-, and tryptophan derivatives were poor choices as acyl group donors. It was found as a rule that those bonds which are known to be most susceptible to thermolysin-catalyzed peptide bond cleavage were also the most suitable points for peptide bond formation. In most cases, the reaction yields could be improved by the addition of sodium chloride or ammonium sulfate, the resulting "salting-out" effect enhancing the *a priori* existing tendency of the products to precipitate.

A detailed study on thermolysin-controlled condensation of N$^\alpha$-Z aspartic acid and phenylalanine methyl or -ethyl esters[101] revealed that the blocking of the hydrophilic β-carboxylate function of aspartic acid with hydrophobic protector groups, as described in the previous report,[100] was not essential for a high coupling efficiency. Z-Asp-OH and H-Phe-OMe or H-Phe-OEt were readily coupled in very good yields.

Oka and Morihara investigated the thermolysin-mediated peptide bond synthesis between Z-Phe-OH and H-Leu-NH₂ under varying experimental conditions.[102] Maximal coupling yields were obtained after 5 hr at pH 7 and 37°C. Under otherwise identical conditions, the catalytic efficiency of α-chymotrypsin was found to be two orders of magnitude lower than that of thermolysin. This outcome does not appear surprising, if one recalls the previously mentioned low activity displayed by chymotrypsin in those cases where carboxyl components having a free α-carboxyl function were used.[19] In the presence of Z-Phe-OH as acyl group donor, the amide, the methylester, the *t*-butylester, and the anilide of leucine were roughly equipotent nucleophilic acceptors. However, a strong dependency of the product yield upon the chemical nature of the individual amino acid residue positioned in the P_1'-site was observed, which is reflected in the following order: H-Leu-NH₂ > H-Phe-NH₂ > H-Ile-NH₂ > H-Val-NH₂ > H-Ala-NH₂. The amino acid amides H-Tyr-NH₂, H-Gly-NH₂, H-Pro-NH₂, and H-D-Leu-NH₂ turned out to be completely useless as amine components. These results, which are largely in agreement with the above mentioned findings of Isowa and Ichikawa[100] cannot be related solely to the degree of solubility of the respective products. For example, Boc-Phe-Tyr-NH₂, a dipeptide which failed to be formed via thermolysin catalysis, certainly would have been less soluble than other dipeptides that were successfully prepared; nevertheless H-Tyr-NH₂ proved to be a poor nucleophilic acceptor. Apparently, tyrosine, though being of hydrophobic nature, glycine-, proline-, and D-amino acid residues in the P_1'-site are incompatible with the geometry of the S_1'-site of the enzyme. The influence of the carboxyl component on thermolysin-catalyzed reactions, using H-Leu-NH₂ as amine component, is described by the following order of efficiency: Z-Phe-OH, Z-Gly-Phe-OH, Z-Gly-Pro-Leu-OH > Z-Ala-OH > Z-Gly-OH > Z-Pro-OH, and Z-Arg-OH. Z-D-Ala-OH, however, was not recognized by thermolysin as acyl group donor.

These and the preceding studies demonstrate that the P_1-specificity of thermolysin with regard to peptide bond formation covers a broad spectrum of amino acid residues. On the other hand, the catalytic action of the enzyme obviously proceeds stereospecifically as far as the amino acids residues both at the carbonyl- and at the imino side of the susceptible peptide bond are concerned.

Experiments designed to study the ability of thermolysin to mediate the synthesis of Z-protected dipeptide amides in aqueous-organic biphasic systems were performed by Kuhl and Jakubke.[103] In the presence of ethyl acetate (>90% v/v) as the organic phase and *tris*-maleate buffer (pH 7.0), Z-Phe-Leu-NH₂ and Z-Leu-Met-NH₂ were obtained in high yields (95 and 80%, respectively). Könnecke et al. who synthesized model peptides[54] and Oyama et al. who prepared a precursor of the synthetic sweetener aspartame,[104] showed that immobilized thermolysin is an equally efficient catalyst for peptide-bond formation as the freely mobile protease. With view to a commercial-scale production of aspartame, Nakanishi et al.[105] attempted the continuous synthesis of its precursor Z-Phe-Asp-OMe. The authors used immobilized thermolysin and conducted the peptide synthesis in an aqueous-organic biphasic system, the aqueous phase of which was largely maintained inside the immobilized protease resin. In the presence of ethyl acetate which was found to be the most suitable organic solvent, a stirred tank reactor was operated for more than 300 hr to give a yield of approximately 90%.

Peptide synthesis in analogy to the well-known solid-phase technology[43] using immobilized, silica bound leucine amide as the amine component and Z-Phe-OH as the carboxyl component in the presence of "mobile" thermolysin was reported by Könnecke et al.[36] However, in contrast to comparable studies on chymotryptic peptide bond formation (yield = 20%), the thermolysin-controlled coupling step gave a markedly improved product yield of 60%.

In summary, the above results of thermolysin-catalyzed syntheses are indicative of a far-reaching applicability of the protease. Isowa and Ichikawa[100] and Isowa et al.[106] examined some additional metalloproteases of microbial origin such as prolisin A, dispase A, and tacynase I, which, in some cases, turned out to be rather useful tools for peptide synthetic preparations. From the point of view of peptide synthesis, however, thermolysin undoubtedly remains the most effective and valuable catalyst of the family of metalloproteases.

REFERENCES

1. **Martinek, K. and Semenov, A. N.,** Enzymes in organic synthesis: physicochemical means of increasing the yields of end product in biocatalysis, *J. Appl. Biochem.,* 3, 93, 1981.
2. **Findeis, J. S. and Whitesides, G. M.,** Enzymic methods in organic synthesis, *Ann. Rep. Med. Chem.,* 19, 263, 1984.
3. **Fruton, J. S.,** The synthesis of peptides, *Adv. Prot. Chem.,* 5, 1, 1949.
4. **Fruton, J. S.,** Proteinase-catalyzed synthesis of peptide bonds, *Adv. Enzymol. Relat. Areas Mol. Biol.,* 53, 239, 1982.
5. **Bergmann, M. and Fruton, J. S.,** On proteolytic enzymes, XIII. Synthetic substrates for chymotrypsin, *J. Biol. Chem.,* 118, 405, 1937.
6. **Bergmann, M. and Fruton, J. S.,** Some synthetic and hydrolytic experiments with chymotrypsin, *J. Biol. Chem.,* 124, 321, 1938.
7. **Davis, N. C.,** Action of proteolytic enzymes on some peptides and derivatives containing histidine, *J. Biol. Chem.,* 223, 935, 1956.
8. **Kaufman, S. and Neurath, H.,** Structural requirements of specific substrates for chymotrypsin. II. An analysis of the contribution of the structural components to enzymatic hydrolysis, *Arch. Biochem. Biophys.,* 21, 437, 1949.
9. **Brenner, M., Müller, H. R., and Pfister, R. W.,** Eine neue enzymatische Peptidsynthese, *Helv. Chim. Acta,* 33, 568, 1950.
10. **Schechter, I. and Berger, A.,** On the size of the active site in proteases. I. Papain, *Biochem. Biophys. Res. Commun.,* 27, 157, 1967.
11. **Kloss, G. and Schröder, E.,** Über enzymatische Verseifung von Peptid-Estern, *Hoppe-Seyler's Z. Physiol. Chem.,* 336, 248, 1964.
12. **Bauer, C.-A.,** Active centers of Streptomyces griseus protease 1, Streptomyces griseus protease 3, and α-chymotrypsin: enzyme-substrate interactions, *Biochemistry,* 17, 375, 1978.
13. **Morihara, K. and Oka, T.,** α-Chymotrypsin as the catalyst for peptide synthesis, *Biochem. J.,* 163, 531, 1977.
14. **Bender, M. L., Clement, G. E., Kezdy, F. J., and Heck, H. D. A.,** The correlation of the pH (pD) dependence and the stepwise mechanism of α-chymotrypsin-catalyzed reactions, *J. Am. Chem. Soc.,* 86, 3680, 1964.
15. **Fastrez, J. and Fersht, A. R.,** Demonstration of the acyl-enzyme mechanism for the hydrolysis of peptides and anilides by chymotrypsin, *Biochemistry,* 12, 2025, 1973.
16. **Morihara, K., Oka, T., and Tsuzuki, H.,** Comparison of α-chymotrypsin and subtilisin BPN': size and specificity of the active site, *Biochem. Biophys. Res. Commun.,* 35, 210, 1969.
17. **Morihara, K. and Oka, T.,** A kinetic investigation of subsites S'_1 and S'_2 in α-chymotrypsin and subtilisin BPN', *Arch. Biochem. Biophys.,* 178, 188, 1977.
18. **Perrin, D. D.,** *Dissociation Constants of Organic Bases in Aqueous Solutions* (and suppl.), Butterworths, London, 1965 and 1972.
19. **Oka, T. and Morihara, M.,** Peptide bond synthesis catalyzed by α-chymotrypsin, *J. Biochem. (Tokyo),* 84, 1277, 1978.
20. **Saltman, R., Vlach, D., and Luisi, P. L.,** Co-oligopeptides of aromatic amino acids and glycine with variable distance between the aromatic residues. VII. Enzymatic synthesis of *N*-protected peptide amides, *Biopolymers,* 16, 631, 1977.
21. **Luisi, P. L., Saltman, R., Vlach, D., and Guarnaccia, R.,** Co-oligopeptides of glycine and aromatic amino acids with variable distance between the aromatic residues. VIII. Enzymatic synthesis of *N*-protected dipeptide esters, *J. Mol. Catalysis,* 2, 133, 1977.
22. **Jones, J. B., Kunitake, T., Niemann, C., and Hein, G. E.,** The primary specificity of α-chymotrypsin. Acylated amino acid esters with normal alkyl side chains, *J. Am. Chem. Soc.,* 87, 1777, 1965.

23. **Bizzozero, S. A., Rovagnati, B. A., and Dutler, H.,** Serine-protease-assisted synthesis of peptide substrates for α-chymotrypsin, *Helv. Chim. Acta,* 65, 1707, 1982.

24. **Bizzozero, S. A., Dutler, H., Franzstack, R., and Rovagnati, B. A.,** Enzyme-catalyzed peptide bond formation: elastase- and δ-chymotrypsin-assisted synthesis of oligopeptides, *Helv. Chim. Acta,* 68, 981, 1985.

25. **Tominaga, M., Pinheiro da Silva Filho, L., Muradian, J., and Seidel, W. F.,** Papain and chymotrypsin as catalysts for peptide synthesis, in *Proc. 20th Symp. Peptide Chemistry 1982,* Sakakibara, S., Ed., Protein Research Foundation, Osaka, Japan, 1983, 271.

26. **Mancheva, I. N., Petkov, D. D., Slavecha, N. N., Videnov, G., Stoev, S. B., and Aleksiev, B. V.,** Enzyme incorporation of cysteine sulphonamide in model peptides, in *Peptides 1984, Proc. 18th Europ. Peptide Symp.,* Ragnarsson, U., Ed., Almquist and Wiksell International, Stockholm, 1984, 189.

27. **Singer, S. J.,** The properties of proteins in nonaqueous solvents, *Adv. Prot. Chem.,* 17, 1, 1962.

28. **Homandberg, G. A., Mattis, J. A., and Laskowksi, M., Jr.,** Synthesis of peptide bonds by proteinases. Addition of cosolvents shifts peptide bond equilibria toward synthesis, *Biochemistry,* 17, 5220, 1978.

29. **Kuhl, P., Könnecke, A., Döring, G., Däumer, H., and Jakubke, H.-D.,** Enzyme-catalyzed peptide synthesis in biphasic aqueous-organic systems, *Tetrahedron Lett.,* 21, 893, 1980.

30. **Semenov, A. N., Berezin, I. V., and Martinek, K.,** Peptide synthesis enzymatically catalyzed in a biphasic system: water - water immiscible organic solvent, *Biotechnol. Bioeng.,* 23, 355, 1981.

31. **Martinek, K. and Semenov, A. N.,** Enzymatic synthesis in biphasic aqueous-organic systems. II. Shift of ionic equilibria, *Biochim. Biophys. Acta,* 658, 90, 1981.

32. **Martinek, K., Semenov, A. N., and Berezin, I. V.,** Enzyme-catalyzed peptide synthesis in a two-phase aqueous-organic system, *Dokl. Akad. Nauk. SSSR,* 254, 121, 1980.

33. **Kuhl, P., Posselt, S., and Jakubke, H.-D.,** α-Chymotrypsinkatalysierte Synthese von Tripeptidamiden im wässrig-organischen Zweiphasensystem, *Pharmazie,* 36, 463, 1981.

34. **Könnecke, A., Bullerjahn, R., and Jakubke, H.-D.,** Peptide synthesis by means of immobilized enzymes. I. Immobilized α-chymotrypsin, *Monatsh. Chem.,* 112, 469, 1981.

35. **Kuhl, P., Walpuski, J., and Jakubke, H.-D.,** Untersuchungen zum Einfluss der Reaktionsbedingungen auf die α-chymotrypsinkatalysierte Peptidsynthese im wässrig-organischen Zweiphasensystem, *Pharmazie,* 37, 766, 1982.

36. **Könnecke, A., Dettlaff, S., and Jakubke, H.-D.,** Model studies on the utility of nucleophiles bound to insoluble supports for enzymatic peptide synthesis, *Monatsh. Chem.,* 113, 331, 1982.

37. **Kuhl, P., Döring, G., and Jakubke, H.-D.,** Proteasekatalysierte (2 + 2)-Segmentkondensation im wässrig-organischen Zweiphasensystem, *Pharmazie,* 38, 371, 1983.

38. **Lüthi, P. and Luisi, P. L.,** Enzymatic synthesis of hydrocarbon-soluble peptides with reverse micelles, *J. Am. Chem. Soc.,* 106, 7285, 1984.

39. **Jakubke, H.-D. and Kuhl, P.,** Proteasen als Biokatalysatoren für die Peptidsynthese, *Pharmazie,* 37, 89, 1982.

40. **Jakubke, H.-D., Kuhl, P., and Könnecke, A.,** Grundprinzipien der proteasekatalysierten Knüpfung der Peptidbindung, *Angew. Chem.,* 97, 79, 1985.

41. **Kuhl, P., Walpuski, J., and Jakubke, H.-D.,** α-Chymotrypsinkatalysierte Peptidsynthesen unter Verwendung des 4-Sulfobenzylrestes als solubisierende Carboxylschutzgruppe, *Pharmazie,* 39, 280, 1984.

42. **Kuhl, P., Walpuski, J., and Jakubke, H.-D.,** Der 2-Thiosulfatoethylrest als solubisierende Schutzgruppe in der enzymatischen Peptidsynthese, *Pharmazie,* 40, 465, 1985.

43. **Merrifield, R. B.,** Solid phase peptide synthesis. I. The synthesis of a tetrapeptide, *J. Am. Chem. Soc.,* 85, 2149, 1963.

44. **Könnecke, A., Pchalek, V., and Jakubke, H.-D.,** Solubisierende Schutzgruppen für enzymatische Peptidsynthesen. Untersuchungen mit Polyoxyethylen-gebundenen Substraten, *Monatsh. Chem.,* 116, 111, 1985.

45. **Mutter, M., Hagenmaier, H., and Bayer, E.,** Eine neue Methode zur Synthese von Polypeptiden, *Angew. Chem.,* 83, 883, 1971.

46. **Huber, R. and Bode, W.,** Structural basis of the activation and action of trypsin, *Acc. Chem. Res.,* 11, 114, 1978.

47. **Walsh, K. A. and Neurath, H.,** Trypsinogen and chymotrypsinogen as homologous proteins, *Proc. Natl. Acad. Sci. U.S.A.,* 52, 884, 1964.

48. **Keil, B.,** The chemistry and structure of peptides and proteins, *Ann. Rev. Biochem.,* 34, 175, 1965.

49. **Keil, B.,** Trypsin, in *The Enzymes,* 3rd Ed., Part III, Boyer, P. D., Ed., Academic Press, New York, 1971, 249.

50. **Oka, T. and Morihara, K.,** Trypsin as a catalyst for peptide synthesis, *J. Biochem. (Tokyo),* 82, 1055, 1977.

51. **Tsuzuki, H., Oka, T., and Morihara, K.,** Coupling between Cbz-Arg-OH and Leu-X catalyzed by trypsin and papain, *J. Biochem. (Tokyo),* 88, 669, 1980.

52. **Inouye, K., Watanabe, K., Morihara, K., Tochino, K., Kanaya, T., Emura, J., and Sakakibara, S.,** Enzyme-assisted semisynthesis of human insulin, *J. Am. Chem. Soc.,* 101, 751, 1979.

53. **Widmer, F., Ohno, M., Smith, M., Nelson, N., and Anfinsen, C. B.,** Enzymatic peptide synthesis, in *Peptides 1982, 17th Europ. Peptide Symp.,* Blaha, K. and Malon, P., Eds., Walter de Gruyter, Berlin, 1983, 375.

54. **Könnecke, A., Hänsler, M., Schellenberger, V., and Jakubke, H.-D.,** Peptidsynthesen mit immobilisierten Enzymen. II. Immobilisiertes Trypsin, Thermolysin und Papain, *Monatsh. Chem.,* 114, 433, 1983.

55. **Morihara, K.,** Comparitive specificity of microbial proteinases, *Adv. Enzymol. Relat. Areas Mol. Biol.,* 41, 179, 1974.

56. **Kraut, J.,** Serine proteases: structure and mechanism of catalysis, *Ann. Rev. Biochem.,* 46, 331, 1977.

57. **Blow, D. M.,** Structure and mechanism of chymotrypsin, *Acc. Chem. Res.,* 9, 145, 1976.

58. **Isowa, Y., Ohmori, M., Ichikawa, T., Kurita, H., Sato, M., and Mori, K.,** The synthesis of peptides by means of proteolytic enzymes, *Bull. Chem. Soc. Jpn.,* 50, 2762, 1977.

59. **Isowa, Y., Ohmori, M., Sato, M., and Mori, K.,** The enzymatic synthesis of protected valine-5 angiotensin amide-1, *Bull. Chem. Soc. Jpn.,* 50, 2766, 1977.

60. **Morihara, K. and Oka, T.,** Peptide bond synthesis catalyzed by subtilisin, papain, and pepsin, *J. Biochem. (Tokyo),* 89, 385, 1981.

61. **Voyushina, T. L., Lyublinskaya, L. A., and Stepanov, V. M.,** Peptide synthesis catalyzed by serine proteinases. Acylpeptide esters as carboxyl components (in Russian), *Bioorg. Khim.,* 11, 738, 1985.

62. **Fruton, J. S.,** Proteinase-catalyzed synthesis of peptide bonds, *Adv. Enzymol. Relat. Areas Mol. Biol.,* 53, 239, 1982.

63. **Gertler, A. and Hofman, T.,** Acetyl-L-alanyl-L-alanyl-L-alanine methyl ester: a new highly specific elastase substrate, *Can. J. Biochem.,* 48, 384, 1970.

64. **Thompson, R. C. and Blout, E. R.,** Dependence of the kinetic parameters for elastase-catalyzed amide hydrolysis on the length of peptide substrates, *Biochemistry,* 12, 57, 1973.

65. **Könnecke, A. and Jakubke, H.-D.,** Versuche zur Anwendung von Thermitase als Katalysator zur Knüpfung der Peptidbindung, *Monatsh. Chem.,* 112, 1099, 1981.

66. **Hayashi, R.,** Carboxypeptidase Y, *Methods Enzymol.,* 45, 568, 1976.

67. **Widmer, F. and Johansen, J. T.,** Enzymatic peptide synthesis. Carboxypeptidase Y catalyzed formation of peptide bonds, *Carlsberg Res. Commun.,* 44, 37, 1979.

68. **Hayashi, R., Bai, Y., and Hata, T.,** Kinetic studies of carboxypeptidase Y. I. Kinetic parameters for the hydrolysis of synthetic substrates, *J. Biochem. (Tokyo),* 77, 69, 1975.

69. **Widmer, F., Breddam, K., and Johansen, J. T.,** Influence of the structure of amine components on carboxypeptidase Y catalyzed amide bond formation, *Carlsberg Res. Commun.,* 46, 97, 1981.

70. **Widmer, F., Breddam, K., and Johansen, J. T.,** Carboxypeptidase Y catalyzed peptide synthesis using amino acid alkyl esters as amine components, *Carlsberg Res. Commun.,* 45, 453, 1980.

71. **Breddam, K., Widmer, F., and Johansen, J. T.,** Amino acid methyl esters as amine components in CPD-Y catalyzed peptide synthesis: control of side reactions, *Carlsberg Res. Commun.,* 48, 231, 1983.

72. **Breddam, K., Widmer, F., and Johansen, J. T.,** Carboxypeptidase Y catalyzed transpeptidations and enzymatic peptide synthesis, *Carlsberg Res. Commun.,* 45, 237, 1980.

73. **Kuhl, P., Zapevalova, N. P., Könnecke, A., and Jakubke, H.-D.,** Model studies on carboxypeptidase Y catalyzed peptide synthesis in an aqueous-organic two-phase system, *Monatsh. Chem.,* 114, 343, 1983.

74. **Bergmann, M. and Fraenkel-Conrat, H.,** The enzymatic synthesis of peptide bonds, *J. Biol. Chem.,* 124, 1, 1938.

75. **Bergmann, M., Zervas, L., and Fruton, J. S.,** On proteolytic enzymes. VI. On the specificity of papain, *J. Biol. Chem.,* 111, 225, 1935.

76. **Mycek, M. J. and Fruton, J. S.,** Specificity of papain-catalyzed transamidation reactions, *J. Biol. Chem.,* 226, 165, 1957.

77. **Brubacher, L. J. and Zaher, M. R.,** A kinetic study of hydrophobic interactions at the S_1 and S_2 sites of papain, *Can. J. Biochem.,* 57, 1064, 1979.

78. **Milne, H. B., Halver, J. E., Ho, D. S., and Mason, M. S.,** The oxidative cleavage of phenylhydrazide groups from carboallyloxy-α-amino acid phenylhydrazides and carboallyloxydipeptide phenylhydrazides, *J. Am. Chem. Soc.,* 79, 637, 1957.

79. **Mitin, Y. V., Zapevalova, N. P., and Gorbunova, E. Y.,** Peptide synthesis catalyzed by papain at alkaline pH values, *Int. J. Peptide Protein Res.,* 23, 528, 1984.

80. **Fox, S. W., Pettinga, C. W., Halverson, J. S., and Wax, H.,** Enzymic synthesis of peptide bonds. II. Preferences of papain within the monoaminomonocarboxylic acid series, *Arch. Biochem.,* 25, 13, 1950.

81. **Fox, S. W. and Pettinga, C. W.,** Enzymic synthesis of peptide bonds. I. Some factors which influence the synthesis of peptide bond as catalyzed by papain, *Arch. Biochem.,* 25, 13, 1950.

82. **Čeřovský, V. and Jošt, K.,** Enzymatically catalyzed synthesis of dipeptides of γ-carboxyl-L-glutamic acid from benzyloxycarbonyl-γ-carboxy-DL-glutamic acid, *Coll. Czech. Chem. Commun.,* 50, 878, 1985.

83. **Wong, C., Cheng, S., and Wang, K.,** Enzymic synthesis of opioid peptides, *Biochim. Biophys. Acta,* 576, 247, 1979.

84. **Döring, G., Kuhl, P., and Jakubke, H.-D.,** Modelluntersuchungen zur Papainkatalysierten Peptidsynthese im wässrig-organischen Zweiphasensystem, *Monatsh. Chem.,* 112, 1165, 1981.

85. **Whitaker, J. R. and Bender, M. L.,** Kinetics of papain-catalyzed hydrolysis of α-*N*-Benzol-L-arginine ethyl ester and α-*N*-Benzoyl-L-arginineamide, *J. Am. Chem. Soc.,* 87, 2728, 1965.

86. **Fruton, J. S.,** The mechanism of the catalytic action of pepsin and related acid proteinases, *Adv. Enzymol. Relat. Areas Mol. Biol.,* 44, 1, 1976.

87. **Fruton, J. S.,** The specificity and mechanism of pepsin action, *Adv. Enzymol. Relat. Areas Mol. Biol.,* 33, 401, 1970.

88. **Florkin, M. and Stotz, E. H.,** The unravelling of biosynthetic pathways, in *Comprehensive Biochemistry,* Vol. 32, Elsevier, Amsterdam, 1977, 307.

89. **Wieland, T., Determann, H., and Albrecht, E.,** Untersuchungen über die Plastein-Reaktion. Isolierung einheitlicher Plastein-Bausteine, *Justus Liebigs Ann. Chem.,* 633, 185, 1960.

90. **Pellegrini, A. and Luisi, P. L.,** Pepsin induced synthesis of peptide bonds, in *Peptides, Proc. 5th Am. Peptide Symp.,* Goodman, M. and Meienhofer, J., Eds., Wiley and Sons, New York, 1977, 556.

91. **Pellegrini, A. and Luisi, P. L.,** Pepsin-catalyzed peptide synthesis, *Biopolymers,* 17, 2573, 1978.

92. **Tseng, M.-J., Wu, S-H., and Wang, K.-T.,** Enzymic synthesis of oligopeptides. VI. The mechanistic features of pepsin-catalyzed peptide synthesis, *Tetrahedron,* 39, 61, 1983.

93. **Isowa, Y.,** Enzymatic peptide synthesis, *Yuki Gasei Kagaku,* 36, 195, 1987.

94. **Kuhl, P., Wilsdorf, A., and Jakubke, H.-D.,** Modelluntersuchungen zur pepsinkatalysierten Peptidsynthese im wässrig-organischen Zweiphasensystem, *Monatsh. Chem.,* 114, 571, 1983.

95. **Latt, S. A., Holmquist, B., and Valley, B. L.,** Thermolysin: a zinc metalloenzyme, *Biochem. Biophys. Res. Commun.,* 37, 333, 1969.

96. **Feder, J., Garret, L. R., and Wildi, B. S.,** Studies on the role of calcium in thermolysin, *Biochemistry,* 10, 4552, 1971.

97. **Morihara, K. and Tsuzuki, H.,** Thermolysin: kinetic study with oligopeptides, *Eur. J. Biochem.,* 15, 374, 1970.

98. **Kester, W. R. and Matthews, B. W.,** Crystallographic study of the binding of dipeptide inhibitors to thermolysin: implications for the mechanism of catalysis, *Biochemistry,* 16, 2506, 1977.

99. **Morihara, K., Tsuzuki, H., and Oka, T.,** Acyl and amino intermediates in reactions catalyzed by thermolysin, *Biochem. Biophys. Res. Commun.,* 84, 95, 1978.

100. **Isowa, Y. and Ichikawa, T.,** Syntheses of *N*-acyl dipeptide derivatives by metalloproteinases, *Bull. Chem. Soc. Jpn.,* 52, 796, 1979.

101. **Isowa, Y., Ohmori, M., Ichikawa, T., Mori, K., Nonaka, Y., Kihara, K., Oyama, K., Satoh, H., and Nishimura, S.,** The thermolysin-catalyzed condensation reactions of *N*-substituted aspartic and glutamic acids with phenylalanine alkyl esters, *Tetrahedron Lett.,* 22, 2611, 1979.

102. **Oka, T. and Morihara, K.,** Peptide bond synthesis catalyzed by thermolysin, *J. Biochem. (Tokyo),* 88, 807, 1980.

103. **Kuhl, P. and Jakubke, H.-D.,** Thermolysin-katalysierte Peptidsynthese im wässrig-organischen Zwei-phasen-system, *Z. Chem.,* 22, 407, 1982.

104. **Oyama, K., Nishimura, S., Nonaka, Y., Kihara, K., and Hashimoto, T.,** Synthesis of an aspartame precursor by immobilized thermolysin in an organic solvent, *J. Org. Chem.,* 46, 5241, 1981.

105. **Nakanishi, K., Kamikubo, T., and Matsuno, R.,** Continous synthesis of *N*-(Benzyloxycarbonyl)-L-aspartyl-L-phenylalanine methyl ester with immobilized thermolysin in an organic solvent, *Biotechnology,* 3, 459, 1985.

106. **Isowa, Y., Ichikawa, T., and Ohmori, M.,** Peptide syntheses with proteinases. Fragment condensation of ZLeuGlnGlyOH or ZGlnGlyOH with HLeuValNH$_2$ using metalloproteinases, *Bull. Chem. Soc. Jpn.,* 51, 271, 1978.

Chapter 8

ENZYMATIC SYNTHESES OF BIOLOGICALLY ACTIVE PEPTIDES

I. INTRODUCTION

The reports on protease-catalyzed syntheses of model peptides examined so far dealt predominantly with the exploration of the proteosynthetic potential and the specificity of certain proteases for the formation of individual peptide bonds. Clearly, the optimal condition under which the enzymatic reactions could take place, had to be elaborated. The extremely valuable informations provided by these systematic studies was essential for broadening the applicability of the enzymatic approach to peptide synthetic chemistry. Due to the model character of these studies the authors' interest was largely aimed at the synthesis of only a single peptide bond via protease catalysis. As a consequence, these syntheses often started from partially protected amino acid residues and did not proceed beyond the level of protected dipeptides. Often where protected peptides served as educts, they had been prepared by chemical means prior to their use in enzymatic synthesis.

In order to establish the enzymatic procedure as an integral part of the peptide synthetic methodology, one must inevitably reach beyond the scope of synthetic model peptides to a more generalized application of the proteases as biocatalysts for peptide synthetic purposes. A good opportunity therefore to demonstrate the capabilities of enzyme-mediated synthesis was the preparation of several biologically active peptides, each peptide bond of which could be prepared via protease catalysis.

II. ENDOGENOUS OPIATE-PEPTIDES

The enzymatic assembly of the endogenous opioid peptides Leu- and Met-enkephalin, the amino acid sequences of which were described by Hughes et al.[1] may represent the first example of the synthetic applicability of proteases to the preparation of naturally occurring peptides.

In the progress of these syntheses, some preliminary difficulties arising principally from protease-inherent proteolytic activities had to be overcome *(vide infra,* Chapter 13). Nevertheless, two alternative synthetic pathways were elaborated which finally permitted the synthesis of the desired opioid peptides whose peptide bonds were prepared exclusively either by papain- or α-chymotrypsin catalysis (Figure 1).[2,3]

Without exception, the peptide-bond-forming steps were solubility-controlled. Thus, two birds were killed with one stone in that the poor solubility of the resulting products firstly induced a favorable shift of the chemical equilibria and second greatly simplified the purification procedures. All but one of these enzymatic coupling steps were performed in monophasic aqueous-organic systems in order to induce a favorable shift of the ionic equilibria and to enhance the solubility of the initial reactants in the respective incubation mixtures.

To ensure the synthesis of well-defined peptides, the application of tempory and semipermanent protecting groups* is, as noted previously (cf. Chapter 6, Section I.B), an indispensable must in chemical peptide synthesis. During enzymatic peptide syntheses, this condition is also valid as far as the temporary protection of α-amino and α-carboxyl groups is concerned. Moreover, the introduction of protecting groups is an essential requirement for thermodynamic reasons *(vide supra,* Chapter 5, Section I.A) and may often serve a

---

* Temporary protecting groups are removed prior to each chain-lengthening step, whereas semipermanent protecting groups removed after completion of the synthesis.

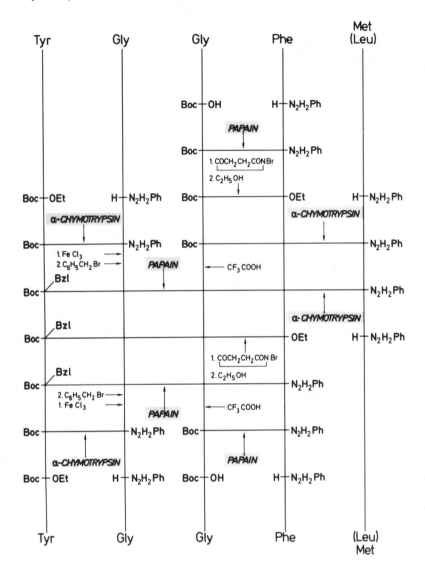

FIGURE 1.   Enzymatic syntheses of Leu- and Met-enkephalin.

further purpose insofar as it ensures a suitable solubility differential between educts and products.

During the syntheses of the enkephalins, N^{α}- and α-carboxyl protection was achieved respectively by t-butyloxycarbonyl- and phenylhydrazide groups. These protector groups could be readily introduced — the phenylhydrazide moiety, for example, via papain-catalysis[4] — and they were also easily and selectively removable under mild conditions. The above combination of protecting groups contributed to the poor solubility of the newly formed products without endangering the solubility of the initial reactants. An esterification of the α-carboxyl groups of the amine components — a popular approach in chemical syntheses — is not advisable for the purposes of enzymic synthesis in view of the esterase activities displayed by many proteases.

Boc-amino acid- or peptide ethyl esters were used as "pre-activated" carboxyl components during α-chymotrypsin catalysis. Esterified acyl group donors were preferred to those having a free α-carboxyl groups because the latter were known to be less efficient in chymotrypsin-controlled peptide bond formation (*vide supra*, Chapter 7, Section I).

Although all the possible dipeptide subunits of both Leu- and Met-enkephalin were readily obtainable as their Boc-dipeptide phenylhydrazides, either by papain- or by chymotrypsin-catalysis, only the synthetic routes outlined in Figure 1 led successfully to the desired opiate-pentapeptides. (A rationale for the failure to realize several ostensibly promising pathways is given below.) The outcome of preliminary experiments suggested an assembly of the target peptides via fragment condensation — as depicted in Figure 1 — in preference to a stepwise incorporation of individual amino acid derivatives into the growing peptide chain.

The last step of the first synthetic route comprised the coupling, via papain-catalysis, of the dipeptide Boc-Tyr(Bzl)-Gly-OH with the tripeptides H-Gly-Phe-Leu-N$_2$H$_2$Ph and H-Gly-Phe-Met-N$_2$H$_2$Ph, respectively, at pH 6.1. These reactions furnished the fully protected Leu- and Met-enkephalin in 82 and 73% yield, respectively. The above acyl component, Boc-Tyr(Bzl)-Gly-OH, was prepared as follows: Boc-Tyr-OEt and H-Gly-N$_2$H$_2$Ph were incubated at pH 10.1 in the presence of α-chymotrypsin, giving Boc-Tyr-Gly-N$_2$H$_2$Ph in 72% yield. The resulting dipeptide was treated successively with ferric chloride to remove the phenylhydrazide moiety and with benzyl bromide to benzylate the phenolic group of the tyrosine unit. This "tactical" *a posteriori* introduction of a benzyl group was in fact indispensable, since the preparation of the desired enkephalin-pentapeptides turned out to be impracticable unless Boc-Tyr-Gly-OH, which had been initially considered as acyl group donor, was replaced by Boc-Tyr(Bzl)-Gly-OH. Evidently, the benzylation of the tyrosine side chain provided the driving force for the papain-catalyzed synthesis by lowering the solubility of the resulting pentapeptides. On the other hand, the benzyl group could not be introduced as integral constituent of a tyrosine derivative, because neither Boc-Tyr(Bzl)-OEt nor Boc-Tyr(Bzl)-OH were capable of being coupled with H-Gly-N$_2$H$_2$Ph via chymotrypsin- or papain-catalysis, respectively.

In the present case, a protecting group originally designed to suppress the reactivity of a side-chain function was harnessed to permit the solubility-controlled synthesis of target peptides. Evidently, in this case the term "protecting group" is misleading, as the benzyl group no longer serves its original purpose, i.e., the elimination of the reactivity of the phenolic function. This matter was closed with the use of chymotrypsin, the regiospecificity of which is strictly confined to the α-amino- or the α-carboxyl group of the substrates.

The N^α-acylated forms of the prospective amine components could be synthesized by incubation of Boc-Gly-OH and H-Phe-N$_2$H$_2$Ph at pH 4.8 in the presence of papain (yield, 80%). The newly formed dipeptide phenylhydrazide Boc-Gly-Phe-N$_2$H$_2$Ph was treated in succession with bromosuccinimide and ethanol[5] to give the corresponding ethyl ester Boc-Gly-Phe-OEt which was then coupled with H-Leu-N$_2$H$_2$Ph or H-Met-N$_2$H$_2$Ph at pH 9.95 (yields: 70 and 61%, respectively).

Concurrently with the development of the synthetic route presented above, an alternative pathway leading to the enzymatic synthesis of the above opiate peptides was also explored. This second enkephalin synthesis commenced with a papain-controlled fragment coupling of Boc-Tyr(Bzl)-Gly-OH and H-Gly-Phe-N$_2$H$_2$Ph at pH 6.0 (Figure 1). In an analogous manner to the procedure described above, the outcoming tetrapeptide phenylhydrazide (yield: 71%) was subsequently rearranged to give the corresponding tetrapeptide ester. Incubation of the latter with H-Leu-N$_2$H$_2$Ph or H-Met-N$_2$H$_2$Ph in the presence of α-chymotrypsin gave the target peptides in 50 and 45% yield, respectively. The above tetrapeptide ethyl ester, Boc-Tyr(Bzl)-Gly-Gly-Phe-OEt, was obtained in rather moderate yields (35%). This shortcoming presumably resulted from the treatment of the precursor tetrapeptide with bromosuccinimide, which is known to be potentially hazardous to tyrosine-containing peptides.[6] The C-terminal elongation of the tetrapeptide ethyl ester by leucine- and methionine phenylhydrazide also did not fully satisfy original expectations. Thus, in the present case the desired products could not be obtained even in the moderate yields noted above unless the enzyme concentration was considerably increased, the incubation period prolonged to 40 hr and the temperature raised to 38°C. In contrast, chymotrypsin-catalyzed syntheses usually proceed at room temperature and were completed after 10 to 15 min. Taking into consideration the results of Peterson et al.[7] obtained for O^n-alkylated tyrosine derivatives, the reduced coupling efficiency may be explained as follows. First, the benzylated tyrosine residue of Boc-Tyr(Bzl)-Gly-Gly-Phe-OEt binds nonproductively to the active site of α-chymotrypsin, thereby inhibiting the formation of the phenylalanyl-enzyme complex which is essential for the synthesis of the desired pentapeptide. In addition, the bulky benzyl ether group may contribute to an unfavorable orientation of the potentially α-chymotrypsin-labile Tyr-Gly-bond, thus preserving its integrity. In view of these considerations, the previously mentioned failure of Boc-Tyr(Bzl)-OEt to react with H-Gly-N_2H_2Ph in the presence of α-chymotrypsin also appears reasonable.

After completion of the syntheses, the fully protected pentapeptides were treated successively with ferric chloride and hydrogen bromide in trifluoroacetic acid, to remove the blocking groups. The resulting free enkephalin peptides were purified to homogeneity and assayed for their biological activities. Indeed the enzymatically prepared enkephalins displayed naloxone-reversible opiate like activities equivalent to their natural analogs, and they were two orders of magnitude more active than a chemically prepared all D-enantiomer of the naturally occurring L-form of Leu-enkephalin.[3]

The result of the enzymatic syntheses of the enkephalins clearly demonstrated the intrinsic capability of the proteases to function as biocatalysts for the preparation of biologically active peptides. Two alternative pathways leading to biologically active opiate peptides have been successfully elucidated although the last-mentioned approach is clearly less efficient than the former. In general, the reaction yields of the individual, protease-controlled peptide-bond-forming step were quite satisfactory and the intermediary products were readily purified due to their extremely low solubility. In addition to the exploitation of the esterase potential of the chymotrypsin, the use of donor esters allows the reaction to proceed very rapidly within a more alkaline environment, under which conditions the proteolytic activity of α-chymotrypsin is greatly reduced while its esterolytic activity is still high.

The incubation period could be limited to a few minutes during chymotrypsin-catalyzed syntheses. Thus, the time spent on synthesis could be markedly reduced in comparison to the time usually spent on chemical syntheses; although the papain-mediated syntheses took roughly the same time as conventional procedures. Certainly the time spent on the total synthesis was less than that required for comparable classical chemical syntheses performed in solution. Since time-consuming purification steps could be largely avoided, the enzymatic syntheses could be completed even more rapidly than comparable solid-phase syntheses.

Papain was exclusively used to catalyze the formation of all peptide bonds during an enzymatic synthesis of Leu-enkephalin reported by Zapevalova et al.[8] The opioid pentpeptide was assembled by a stepwise mode in conjunction with a C → N coupling schema, and monodirectional chain growth proceeded via repeated incorporation of successive protected amino acid units. Z-protected amino acid methyl esters functioned as carboxyl components while leucine *t*-butyl ester and, with growing chain length, free di-, tri-, and tetrapeptides, respectively, served as amine components. In order to prevent secondary hydrolysis of preexisting peptide bonds, the papain-mediated coupling steps were conducted within an alkaline environment (pH = 8.1 to 9.6) in analogy to a procedure described by Mitin et al.[9] The product yields ranged from 74 to 79%; the final coupling reaction, however, gave the target pentapeptide in a rather low yield (23%).

A different strategy for the enzymatic synthesis of Met-enkephalin was developed by Widmer et al.[10] This approach was based principally on carboxypeptidase Y-catalyzed condensation- and desamidation steps (Figure 2). CPD-Y is an exopeptidase which specifically acts on the carboxyl terminus of a given substrate. Consequently, the enzymatic synthesis

65

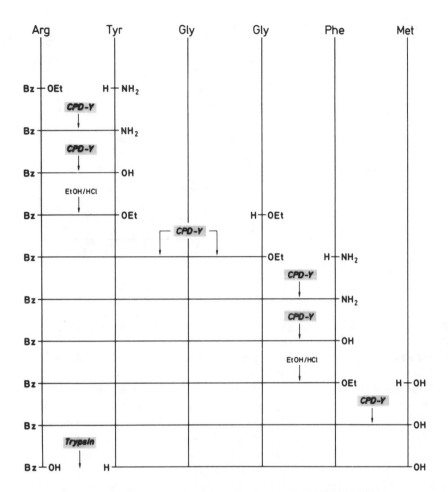

FIGURE 2. Carboxypeptidase Y-catalyzed synthesis of Met-enkaphalin.

proceeds by the consecutive addition of the individual amino acid derivatives to the C-terminal end of the growing peptide chain. Due to the ability of CPD-Y, to catalyze in an orderly fashion the dimerization of H-Gly-OEt,[11] the dipeptide H-Gly-Gly-OEt could be incorporated through a single coupling step (Figure 2). The semipermanent N^α-"protection" of the amino-terminal tyrosine residue was provided by the trypsin-labile benzoylarginine moiety, which could be finally removed by trypsin-catalysis. N^α-benzoylated amino acid or peptide ethyl esters served as acyl group donors, whereas H-Tyr-NH₂, H-Gly-OEt, H-Phe-NH₂, and H-Met-OH were used in the given order as amine components. The reactions were performed in aqueous solutions in the pH range of 8.0 to 9.6 and coupling yields were between 60 and 90%. Prior to elongation of the peptide chain the amide groups were removed by CPD-Y catalysis and subsequently replaced by an ethyl ester group. The CPD-Y approach to enkephalin synthesis took advantage of a salient feature of the peptidase, namely that it lacks the capacity to cleave internal peptide bonds. However, this picture is darkened by transpeptidation phenomena,[12] which may occur during each elongation step. As a consequence, the last-formed peptide bond continuously runs the risk of being hydrolyzed during the subsequent condensation step.

In connection with the design and synthesis of an opiate-receptor mimetic peptide a series of chemically modified Leu-enkephalin peptides were required to explore the binding characteristics of the artificial receptor molecule.[13] The synthesis of three analogues of Leu-enkephalin in which the N-terminal tyrosine moiety was replaced by phenylalanine, phloretic

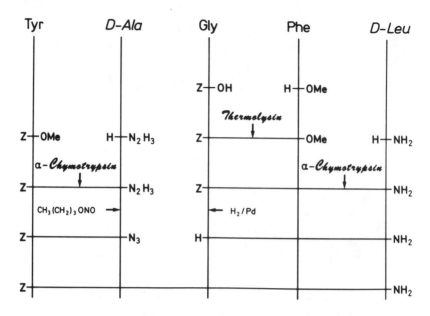

FIGURE 3.   Enzymo-chemical synthesis of (D-Ala$_2$, D-Leu$_5$)-enkephalin amide

acid, or hydrocinnamic acid to give, respectively, (Phe$_1$)-, (Des-NH$_2$-Tyr$_1$)-, and (Des-NH$_2$-Phe$_1$)-Leu-enkephalin, was accomplished by enzymic and mixed enzymochemical procedures. The first-mentioned peptide, having a phenylalanine residue in place of the original tyrosine unit was prepared in an analogous route to the first synthetic route leading to Leu-enkephalin mentioned above.[3] (Des-NH$_2$-Tyr$_1$)- and (Des-NH$_2$-Phe$_1$)-Leu-enkephalin were synthesized from the enzymatically prepared tripeptide H-Gly-Phe-Leu-N$_2$H$_2$Ph[3] by a stepwise approach using DCC and HOBt as condensing agents. The coupling of Boc-Gly-OH and the subsequent deacylation step were followed by acylation of the resulting tetrapeptide with 3-(4-hydroxyphenyl)propionic acid (phloretic acid) and 3-phenylpropionic acid (hydrocinnamic acid), respectively.

The enkephalins commonly undergo rapid enzymatic degradation in vivo, and therefore they only exert short-lasting biological activities. Within the central nervous system there exist a number of specific peptidases the activities of which are probably responsible for the inactivation of the endogenous opiate peptides.[14] However, by D-amino acid substitution in certain positions of the molecule the enkephalins can be made largely resistant to peptidase activities. As a result, these "chirally mutated" opiate peptides can elicit long-lasting pharmacological effects. While the first diastereomeric analogs of the naturally occurring enkephalins originated from attempts to stabilize the peptide molecule against proteolytic cleavage, more recently analogs have often been designed with regard to their selectivity for individual opiate-receptor subclasses.[15] The currently used opiate-ligand (D-Ala$_2$-D-Leu$_5$)-enkephalin (DADLE) exhibits a marked preference for opiate receptors of the δ-subtype.[16] In many instances the use of the above or other diastereomeric peptides for this purpose requires compounds of highest possible optical purity. Consequently, the formulation of an enzymochemical approach to the incorporation of D-amino acids into Leu-enkephalin amide analogs, merits some attention. The successful application of this methodology has been reported by Stoineva and Petkov.[17] The synthetic pathway leading to (D-Ala$_2$ and D-Leu$_5$) enkephalin amide is outlined in Figure 3. In this case, the desired peptide was finally assembled via azide fragment coupling of the dipeptide Z-Tyr-D-Ala-NHNH$_2$ and the tripeptide H-Gly-Phe-D-Leu-NH$_2$. The former was prepared at pH 9.3 in 80% yield from Z-Tyr-OMe and H-D-Ala-NHNH$_2$ via chymotrypsin- catalysis. The N$^\alpha$-acylated form of the latter was obtained by a thermolysin-mediated condensation of Z-Gly-OH and H-Phe-OMe (yield, 65%: pH, 6.8) followed by the chymotryptic coupling of the resulting dipeptide with H-D-Leu-NH₂ (yield, 85%: pH, 9.3). As compared to their L-enantiomeric antipodes, the D-amino acid derivatives display a rather low nucleophilic reactivity in the presence of chymotrypsin (vide supra, Chapter 7, Section I). This drawback was overcome by performing the enzymatic peptide bond formation using an iterative procedure in a nucleophile pool[18] (vide infra, Chapter 13) which enabled the authors to efficiently incorporate the D-amino acid by chymotryptic catalysis. Additionally, two other optical isomers of Leu-enkephalin were prepared in an analogous manner to the synthetic route described above (Figure 3), namely (D-Ala₂-Leu₅)- and (Ala₂-D-Leu₅)-enkephalin. The results of in vitro tests (inhibition of electrically induced contraction of the guinea pig ileum) revealed that the incorporation of D-amino acids increased the inhibitory potency relative to the parent compount of (D-Ala₂, Leu₅)- and (D-Ala₂, -D-Leu₅)-enkephalin by a factor of 540 and 19,000, respectively, whereas a drop in potency (by a factor of 3) was observed for (D-Leu₅)-enkephalin.

According to Morihara et al.,[19] the "proteosynthetic" specificity of chymotrypsin as related to the P_1'-site of its substrates is not strictly confined to amino acid residues of the "natural" L-configuration. Given this finding (cf. Chapter 7, Section I), the successful incorporation of D-amino acids does not come as a surprise. Opposite to this, papain which exhibits indeed a broad secondary specificity with regard to the P_1-site of its substrates, seems completely to reject peptidic compounds having a D-amino acid residue in this position. An attempt, in the presence of papain, to couple the acyl components Boc-Tyr(Bzl)-D-Ala-OH or Boc-Tyr-D-Ala-OH (the Tyr-D-Ala bond of both peptides had been formed via chymotrypsin-catalysis) and the amine component H-Gly-Phe-Leu-N₂H₂Ph (cf. Reference 3) did not give the desired pentapeptide.[20] Although having an amino acid and an amino acid derivative, respectively, in the P_2-position both of which are compatible with the primary specificity of papain,[3,21] the above acyl components were not accepted as papain substrates. Presumably, a D-alanine residue in the P_1-site does not fit the corresponding S_1-position of the active site of papain.

Based on the practical experience gained during the enzymatic syntheses of the enkephalins,[3] a so-called "big" Leu-enkephalin has been prepared.[22] Again, all the peptide bonds of this octapeptide were formed via protease-catalysis (Figure 4). The primary structure of the enkephalin-related peptide the occurrence of which in the porcine hypothalamus and the pituitary of rats has been reported by Minamino et al.[23] and Seizinger et al.,[24] respectively, is identical with the subsequence (1 to 8) of dynorphin and reads as follows:

H-Tyr-Gly-Gly-Phe-Leu-Arg-Arg-Ile-OH.

The isolation of the extraordinary potent opioid peptide dynorphin was described by Goldstein et al.,[25] and the complete primary structure of the heptadecapeptide was elucidated by Tachibana et al.[26] In a systematic study using successively truncated dynorphin peptides Chavkin and Goldstein[27] showed that the biological activity of the N-terminal octapeptide of dynorphin was about 15 times that of Leu-enkephalin. The enzymatic preparation of dynorphin₁₋₈ was performed firstly to provide a deeper insight into protease-catalyzed peptide synthesis, but also to tackle one of the major problems of peptide synthetic chemistry: namely, the synthesis of arginine-peptides.[28-30] Despite the strongly basic character of the δ-guanidine group of arginine (pK_a = 12.5), which is therefore generally protonated under conditions prevailing in peptide synthesis, the low solubility in organic solvents of charged arginine derivatives and their tendency to form lactams during the activation of the carboxyl groups, often necessitates N^G-protection. However, the nucleophilicity of the δ-guanidine function, even in the protected form, is high enough to permit an intramolecular reaction of the carbonyl with the vicinal guanidine nitrogen, resulting in cyclization to piperidones.[29]

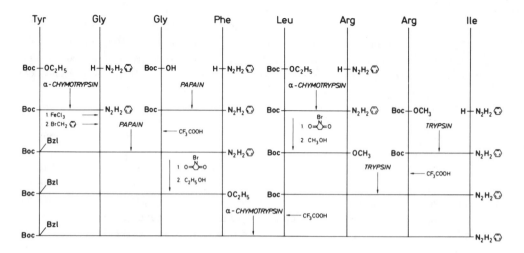

FIGURE 4.   Protease-catalyzed synthesis of dynorphin$_{1-8}$.

Consequently, complete protection to fully suppress the basicity of the δ-guanidine group requires N^ω, N$^{ω'}$-diacylation by bulky protecting groups.[29] However, the advantages of this kind of protection are to some extent nullified by steric hindrances during the coupling steps and by difficulties arising during the selective removal of N^α-protecting groups.[31]

For these reasons, biologically active oligopeptides containing arginine residues or even Arg-Arg subsequences are difficult to prepare. The enzymatic synthesis of dynorphin, which actually contains an Arg-Arg-subunit, was used to develop an alternative approach to the preparation of arginyl peptides relying on the inherent capacity of trypsin to catalyze peptide bond formation when the amino acid which contributes the carbonyl portion of the bond to be formed is an arginine or a lysine residue. The regio- and stereospecificity of tryptic action prevents the occurrence of undesired side reactions, such as lactam formation and racemization, thus enabling the accurate introduction of arginine residues into growing peptide chains.

The synthetic pathway leading to the dynorphin-related octapeptide is illustrated in Figure 4. This work employed a strategy of functional group protection similar to that which was used successfully during the enzymatic syntheses of the enkephalins. The results of these syntheses were also instructive in that they suggested the possibility of using esterified acyl group donors for tryptic and chymotryptic peptide bond formation, in order to take advantage of the esterase activities displayed by the two serine proteases. Furthermore, the amine components were added in a molar excess, to improve the degree of peptide synthesis (*vide supra,* Chapter 5, Section II.C and D).

The protected dynorphin$_{1-8}$, Boc-Tyr(Bzl)-Gly-Gly-Phe-Leu-Arg-Arg-Ile-N$_2$H$_2$Ph, was obtained by α-chymotrypsin-controlled coupling of Boc-Tyr(Bzl)-Gly-Gly-Phe-OEt and H-Leu-Arg-Arg-Ile-N$_2$H$_2$Ph in 52% yield (Figure 4). Although the Leu-Arg bond should in principle be chymotrypsin labile, indeed it was prepared via α-chymotrypsin catalysis, there was no evidence of proteolytic cleavage of this peptide bond. This finding is reasonable when one takes into account the prevailing alkaline reaction conditions (pH 10.2), which greatly reduce the peptidase activity of the enzyme while maintaining its esterase potential. This is consistent with the observation of Kangawa et al.[32] who found that α-chymotrypsin cleaved primarily the Phe-Leu linkage whereas the Leu-Arg bond is hydrolyzed to a much lesser extent.

The protected N-terminal tetrapeptide fragment of the dynorphin octapeptide is identical to one of the enkephalin subunits (cf. Figure 1) and its assembly will not be a subject to further discussion. The acylated form of the above amine component, Boc-Leu-Arg-Arg-Ile-N₂H₂Ph, was prepared in 64% yield by trypsin-mediated condensation of the dipeptides of Boc-Leu-Arg-OMe and H-Arg-Ile-N₂H₂Ph. In analogy to the just mentioned chymotrypsin-resistant Leu-Arg bond, the potentially trypsin-susceptible Arg-Ile bond was scarcely affected during this reaction. In this case also the inertness of this bond is probably due to the relatively high pH (10.4) of the reaction medium, at which tryptic hydrolysis is largely suppressed. This observation agrees with several reports on the tryptic hydrolysis of dynorphin-related peptides the Arg-Ile bond of which was less sensitive than the Arg-Arg linkage.[23,26]

The above acyl component Boc-Leu-Arg-OMe was obtained by treating Boc-Leu-Arg-N₂H₂Ph successively with N-bromosuccinimide and methanol. Meanwhile Boc-Leu-Arg-N₂H₂Ph was prepared from Boc-Leu-OEt and H-Arg-N₂H₂Ph in the presence of α-chymotrypsin (70%). Boc-Arg-Ile-N₂H₂Ph, the acylated form of the prospective amine component, was obtained in 65% yield by reacting Boc-Arg-OMe and H-Ile-N₂H₂Ph in the presence of trypsin.

After completion of the synthesis, the protecting groups were removed to release the free octapeptide which was subsequently purified to homogeneity. The enzymatically prepared dynorphin-fragment was 11 times more potent than Met-enkephalin in an opiate receptor binding assay.

Despite the analogies between chymotryptic and tryptic peptide syntheses, the pronounced bias of trypsin in favor of basic amino acid residues located in the P_1-site of its substrates, causes one crucial difference. In contrast to the above chymotryptic syntheses, the trypsin-catalyzed syntheses were not solubility controlled, because the resulting arginyl-peptides were quite soluble in the aqueous environment. Consequently, peptide bond formation via trypsin-catalysis had to be performed under kinetic control. In actual fact the reactions could be terminated well before secondary hydrolysis of the products came into being, since optimal product yields were obtained after incubation periods of only 2 and 5 min, respectively. The first traces of secondary hydrolytic products were observed only when the reaction was allowed to proceed for more than 50 min.

III. CHOLECYSTOKININ AND RELATED PEPTIDES

The enzymatic synthesis of the C-terminal octapeptide amide of cholecystokinin (CCK-8) represents a further example of the applicability of the proteases as catalysts in peptide synthetic chemistry. The primary structure of the cholecystokinin fragment comprises members of nearly all classes of amino acids; there are present: hydrophilic and hydrophobic, aliphatic and aromatic, charged and uncharged polar amino acid residues. This study is chiefly aimed at preparing the CCK-octapeptide amine as far as possible by enzymatic peptide bond formation. CCK-8, the amino acid sequence of which reads as follows:

H-Asp-Tyr (SO₃H)-Met-Gly-Trp-Met-Asp-Phe-NH₂

elicits the cholecystokinetic and pancreozyminic activities of the entire tritriacontapeptide hormone.[33]

The octapeptide was finally assembled by the chemical condensation of two tetrapeptide segments which had been synthesized through the concerted actions of several proteases of different specificity, namely by papain-, thermolysin-, and α-chymotrypsin catalysis.[34] The design of the synthetic strategy, as outlined in Figure 5, had to take account of several features of the peptide sequence of interest. First, the CCK-8 molecule possesses a central glycine residue, a feature which permits the formation of the Gly-Trp bond by chemical means without endangering the chiral integrity of the resulting octapeptide. Second, the

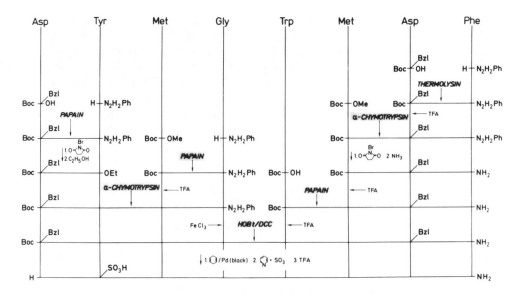

FIGURE 5.   Synthesis of the C-terminal octapeptide amide of cholecystokinin.

results of preliminary studies indicated that fragments containing more than four amino acid units were not attainable exclusively by enzymatic means. Furthermore, the N- and C-terminal tetrapeptides (Figure 5) could not be coupled by either papain-, thermolysin-, ficin-, or bromelain catalysis, because at least one of the tetrapeptides was subject to proteolytic degradation before the synthesis of the desired peptide bond could be completed.

In this strategy the prospective acyl-group donor Boc-Asp(OBzl)-Tyr-Met-Gly-N_2H_2Ph was synthesized via α-chymotrypsin-catalyzed fragment coupling of Boc-Asp(OBzl)-Tyr-OEt and H-Met-Gly-N_2H_2Ph at pH 10.1 (yield: 48%).

The acyl-group donor ester Boc-Asp (OBz1)-Tyr-OEt in turn was obtained by incubation of Boc-Asp(OBz1)-OH and H-Tyr-N_2H_2Ph with papain, following which the resulting dipeptide Boc-Asp(OBzl)-Tyr-N_2H_2Ph (yield: 33%) was treated successively with bromosuccinimide and ethanol,[5] in order to replace its phenylhydrazide group by an ethyl ester to give Boc-Asp(OBzl)-Tyr-OEt. Although the thermolysin-catalyzed coupling of Boc-Asp(OBzl)-OH and H-Phe-N_2H_2Ph was quite efficient *(vide infra)*, the desired dipeptide Boc-Asp(OBzl)-Tyr-N_2H_2Ph could not be prepared from Boc-Asp(OBzl)-OH and H-Phe-N_2H_2Ph in the presence of thermolysin. This failure is explicable by the already mentioned ''aversion'' of thermolysin for tyrosine residues in the P'_1-site. Boc-Met-Gly-N_2H_2Ph, the deacylated precursor of the above acceptor nucleophile, was obtained by papain-mediated coupling of either Boc-Met-OMe or Boc-Met-OH with H-Gly-N_2H_2Ph. Considering the improved coupling yield (61 vs. 51%), Boc-Met-OMe was preferred as acyl group donor over Boc-Met-OH.

The original attempt to prepare Boc-Asp(OBzl)-Tyr-Met-Gly-N_2H_2Ph by protease-catalyzed stepwise incorporation of successive amino acid residue was a failure. Although the tripeptide Boc-Tyr-Met-Gly-H_2H_2Ph was accessible via chymotryptic coupling of Boc-Tyr-OEt and H-Met-Gly-N_2H_2Ph, the enzymatic elongation of the deacylated dipeptide H-Tyr-Met-Gly-N_2H_2Ph by Boc-Asp(OBzl)-OH could not be achieved. Neither thermolysin nor papain were able to catalyze the synthesis of the desired tripeptide. In the presence of papain, the Met-Gly-bond of the amino component was merely split. To avoid proteolytic cleavage of pre-existing peptide bonds, the synthetic pathway described in Figure 5 was chosen: i.e., α-chymotrypsin-mediated fragment condensation of two dipeptides, separately prepared by papain catalysis, which enabled the successful synthesis of the target tetrapeptide.

The C-terminal tetrapeptide Boc-Trp-Met-Asp(OBzl)-Phe-NH$_2$ was prepared by a stepwise procedure through the concerted action of three proteases of different specificities (Figure 5). Boc-Asp(OBzl)-Phe-N$_2$H$_2$Ph was prepared in 52% yield by incubating Boc-Asp(OBzl)-OH and H-Phe-N$_2$H$_2$Ph in the presence of thermolysin. Although the dipeptide amide, Boc-Asp(OBzl)-Phe-NH$_2$, was more easily available by thermolysin-catalysis than the above phenylhydrazide-protected dipeptide, the latter was chosen as prospective amine component to enable the synthesis of the Met-Asp bond which could not be formed via incubation of Boc-Met-OMe or Boc-Met-OH and H-Asp(OBzl)-Phe-NH$_2$ in the presence of α-chymotrypsin or papain. However, the desired chain elongation to the tripeptide Boc-Met-Asp(OBz)-Phe-N$_2$H$_2$Ph could be achieved by α-chymotrypsin-controlled coupling of Boc-Met-OMe and H-Asp(OBzl)-Phe-N$_2$H$_2$Ph (yield 47%). The phenylhydrazide-protection of the tripeptide was then replaced by an amide group by successive treatment with bromosuccinimide and ammonia (yield: 54%). This rearrangement was performed prior to the incorporation of the N-terminal tryptophan residue into the growing peptide chain so as to prevent any undesired side reaction of N-bromosuccinimidine with the indole moiety of tryptophan.[6] Boc-Trp-Met-Asp(OBzl)-Phe-NH$_2$ was finally obtained by acylation of H-Met-Asp(OBzl)-Phe-NH$_2$ with Boc-Trp-OH in the presence of papain (yield: 54%). Despite the preference of chymotrypsin for aromatic amino acid residues in the P$_1$-position of its substrates, the tetrapeptide could not be prepared by α-chymotrypsin-catalyzed coupling of Boc-Trp-OEt and H-Met-Asp(OBzl)-Phe-NH$_2$, because the donor ester could not be cleaved by the enzyme.

As already mentioned, all attempts, via protease catalysis, to couple Boc-Asp(OBzl)-Tyr-Met-Gly-OH and H-Trp-Met-Asp(OBzl)-Phe-N$_2$H$_2$Ph to yield the desired octapeptide amide were unsuccessful. Consequently, it was finally assembled in the presence of the chemical condensing agents dicyclohexylcarbodiimide and 1-hydroxybenzotriazole.

After removal of the protecting groups and sulfation of the phenolic group of the tyrosine residue, the synthetic octapeptide was purified to homogeneity. The enzymatically prepared cholecystokinin fragment stimulated the secretion of amylase by pancreatic cells in an analogous fashion to its naturally occurring counterpart.

Except for the substitution of a threonine for a methionine at the third amino acid position, CCK-8 is identical to the C-terminal octapeptide of cerulein[35] which exhibits similar biological functions to cholecystokinin.[36] In order to test the applicability of the synthetic pathway leading to the N-terminal tetrapeptide of CCK-8 (cf. Figure 5) to the synthesis of the corresponding cerulein fragment, the tetrapeptide Boc-Asp(OBzl)-Tyr-Thr(Bzl)-Gly-N$_2$H$_2$Ph was synthesized by α-chymotrypsin-catalyzed condensation of Boc-Asp(OBzl)-Tyr-OEt with H-Thr(Bzl)-Gly-N$_2$H$_2$Ph (yield: 40%) (Figure 6).[34] The dipeptide Boc-Thr(Bzl)-Gly-N$_2$H$_2$Ph, the acylated form of the prospective amine component, was prepared in 43% yield by incubation of Boc-Thr(Bzl)-OH and H-Gly-N$_2$H$_2$Ph in the presence of papain. The threonine benzyl ether was then incorporated into the tetrapeptide so as to permit the eventual sulfation of the phenolic function of tyrosine without affecting the alcoholic moiety of threonine.

The total synthesis of a protected cerulein derivative was reported by Takai et al.[37] The assembly of the decapeptide involved enzymatic as well as chemical coupling steps. Based on the known preference of chymotrypsin and the acid proteases[38] for aromatic amino acid residues in the P$_1$-site of their substrates, the authors decided to separately prepare three subunits prior to the final assembly of the entire molecule via fragment condensation in the presence of the above-mentioned proteases. These peptide segments corresponded to the positions 1 to 4, 5 to 7, and 8 to 10 of the primary structure of cerulein (cf. Figure 7). The synthesis of the protected cerulein$_{1-4}$ was carried out in the following way (Figure 7); since the dipeptide Z-pGlu-Gln-OH could not be obtained via protease-catalysis, it was prepared instead by chemical means. The deacylated form of the dipeptide Z(OMe)-Asp(OBzl)-Tyr-OMe, which had been synthesized from Z(OMe)-Asp(OBzl)-OH and H-Tyr-OMe in the

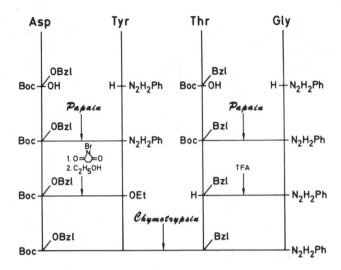

FIGURE 6.   Enzymatic synthesis of the cerulein$_{3-6}$ fragment.

presence of thermolysin, was chemically coupled with the above dipeptide to give the targeted tetrapeptide. (Preceeding attempts to prepare the Gln-Asp bond between these dipeptides in the presence of proteases had failed.) Conversely, the condensation steps leading to the protected cerulein$_{5-7}$ fragment were carried out in the presence of proteases (Figure 7). The dipeptide Z(OMe)-Gly-Trp-OMe was obtained from its constituents by papain catalysis. The elongation step leading to the tripeptide Boc-Thr-Gly-Trp-OMe was mediated either by papain or subtilisin BPN'. The latter was preferred as catalyst, because the product yield of this reaction significantly exceeded that of the papain-catalyzed reaction. The third fragment, cerulein$_{8-10}$, was prepared by the synthesis of the dipeptide Z(OMe)-Asp(OBzl)-Phe-NH$_2$ from Z(OMe)-Asp(OBzl)-OH and H-Phe-NH$_2$ in the presence of the metalloprotease prolisin. The N-terminal methionine residue was then incorporated via papain-catalysis, as Z(OMe)- or Boc-Met-OH. In view of the resulting improved coupling efficiency, the Z(OMe)-protected derivative was preferred over the Boc-derivative. Condensation of the above tripeptide subunits in the presence of the acid protease No. 306 (a fungal protease which is isolated from *Acyndrium* sp.) gave the desired hexapeptide fragment$_{5-10}$ (Figure 7). This condensation step necessitated the formation of the Trp$_7$-Met$_8$ peptide bond, which could also be synthesized in good yields in the presence of the acid proteases pepsin and proctase.[39] [However, in contrast to the acid protease-catalyzed reaction, the hexapeptides fragment$_{5-10}$ resulting from the pepsin- and proctase catalysis lacked the benzyl ether protection of the Thr$_5$ residue (cf. Figure 7).]

The synthesis of the entire, protected cerulein was finally achieved by chymotrypsin-mediated condensation of the subunits 1 to 4 and 5 to 10 (Figure 7) which required the formation of the Thr$_4$-Thr$_5$ peptide bond. This peptide linkage could also be established in the presence of the acid protease No. 306: however, the product yield was rather low (4%). The synthetic cerulein was shown to demonstrate biological activity after sulfation of the Tyr$_4$ side chain and removal of the protecting groups.

IV. MELANOTROPIN-, ELEDOISIN-, AND "EPIDERMAL GROWTH FACTOR" FRAGMENTS

The efficiency of the enzymatic approach to peptide synthesis particularly in the preparation of complex peptides was further emphasized by the protease-catalyzed synthesis of several melanocyte-stimulating hormone (MSH) fragments.[40] The presumptive "active core"[41] of

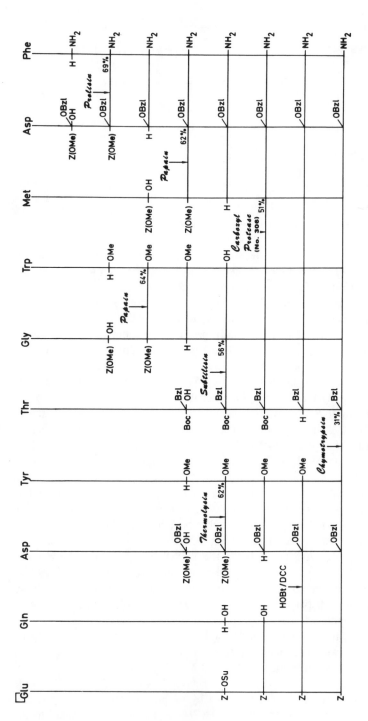

FIGURE 7. Total synthesis of cerulein.

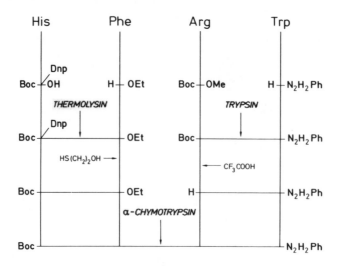

FIGURE 8.   Synthesis of the γ-MSH$_{9-12}$ fragment (bovine).

the adrenocorticotropic hormone (ACTH), the sequence H-His-Phe-Arg-Trp-OH, which is common to all MSH- and ACTH-peptides studied so far,[42] was reported to represent the minimal structural requirement for MSH- and ACTH-like activities.[43] In the course of several conventional syntheses, the presence of histidine, arginine, and tryptophan residues in this biologically essential tetrapeptide gave rise to a variety of problems.[43-45]

Nevertheless, the "active core" could be readily prepared by using three proteases of different specificities (Figure 8). The tetrapeptide Boc-His-Phe-Arg-Trp-N$_2$H$_2$Ph was assembled by fragment condensation of Boc-His-Phe-OEt and H-Arg-Trp-N$_2$H$_2$Ph in the presence of α-chymotrypsin (yield: 50%) (Figure 8).[40] The acyl-group donor Boc-His-Phe-OEt was obtained from Boc-His(Dnp)-OH and H-Phe-OEt via thermolysin catalysis (yield: 74%) followed subsequently by thiolytic removal of the N^{im}-Dnp group using 2-mercaptoethanol (90%). The temporary N^{im}-2,4-dinitrophenyl-protection was found to be essential because Boc-His-OH turned out to be a poor substrate for thermolysin and, furthermore, the N^{im}-protected dipeptide Boc-His(Dnp)-Phe-OEt was inefficient acyl-group donor during the α-chymotrypsin-catalyzed reaction with H-Arg-Phe-N$_2$H$_2$Ph. Although Boc-His-Phe-N$_2$H$_2$Ph was available via papain catalysis (approximate yield: 40%), the corresponding dipeptide ester Boc-His-Phe-OEt could not be prepared to any significant extent in the presence of this enzyme. This failure most probably resulted from a papain-induced oligomerization of H-Phe-OEt; a similar phenomenon had already been reported by Tauber,[46] and Anderson and Luisi[47] for H-Phe-OMe. Boc-Arg-Trp-N$_2$H$_2$Ph, the N-acylated from of the above amine component was obtained in 72% yield by a trypsin-controlled reaction between Boc-Arg-OMe and H-Trp-N$_2$H$_2$Ph. After removal of the protecting groups and purification to homogeneity, the free tetrapeptide displayed lipolytic activities comparable with those of conventionally prepared tetrapeptides.[43]

In addition to the above, the two tetrapeptide fragments of bovine γ-MSH flanking on both sides the "active core" were enzymatically prepared (Figures 9 and 10).[40] The original design for the synthesis of the tetrapeptide Boc-Tyr-Val-Met-Gly-N$_2$H$_2$Ph, involving a stepwise, monodirectional chain growth, was discarded for the following reasons: the dipeptide Boc-Met-Gly-N$_2$H$_2$Ph, which could be obtained both by papain- and α-chymotrypsin-catalyzed coupling of Boc-Met-OMe and H-Gly-N$_2$H$_2$Ph,[34,40] could not be elongated to the tripeptide Boc-Val-Met-Gly-N$_2$H$_2$Ph. The failure to prepare this tripeptide from Boc-Val-OH and H-Met-Gly-N$_2$H$_2$Ph in the presence of thermolysin was not surprising given, the observation of Isowa and Ichikawa,[48] that valine derivatives were inadequate acyl components

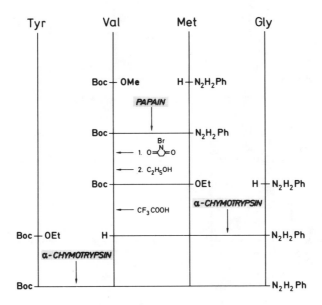

FIGURE 9. Synthesis of the γ-MSH$_{5-8}$ fragment (bovine).

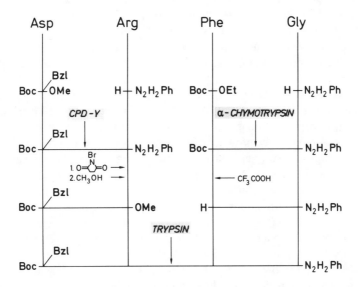

FIGURE 10. Synthesis of the γ-MSH$_{13-16}$ fragment (bovine).

for thermolysin catalysis. A further attempt to obtain the desired tripeptide from Boc-Val-OMe and H-Met-Gly-N$_2$H$_2$Ph via papain catalysis resulted in the proteolytic cleavage of the Met-Gly bond of the amine component.

The originally designed pathway was therefore replaced by a bidirectional, synthetic route, which finally led to the target peptide (Figure 9). The preparation of Boc-Tyr-Val-Met-Gly-N$_2$H$_2$Ph was started with the papain-catalyzed synthesis of Boc-Val-Met-N$_2$H$_2$Ph from Boc-Val-OMe and H-Met-N$_2$H$_2$Ph (52%), followed by the conversion of the dipeptide phenylhydrazide to the corresponding dipeptide methyl ester (55%) which in turn could be reduced with H-Gly-N$_2$H$_2$Ph in the presence of α-chymotrypsin to give the desired tripeptide Boc-Val-Met-Gly-N$_2$H$_2$Ph an 81% yield.

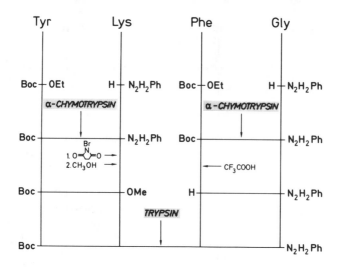

FIGURE 11.   Synthesis of the β-MSH$_{5\text{-}8}$ fragment (dogfish).

The tetrapeptide Boc-Tyr-Val-Met-Gly-N$_2$H$_2$Ph was finally prepared from Boc-Tyr-OEt and H-Val-Met-Gly-N$_2$H$_2$Ph in 63% yield via α-chymotrypsin catalysis. The latter coupling step was successful, the reaction being completed within 15 min without affecting the chymotrypsin-sensitive Met-Gly bond, the formation of which required a considerably prolonged incubation period (2 hr). Obviously, chymotrypsin exhibits a striking preference for tyrosyl- over methionyl derivatives positioned in the P$_1$-site of its substrates. This feature enables the selective formation of more than one peptide bond within a single peptide with the same protease.

The tetrapeptide Boc-Asp(OBzl)-Arg-Phe-Gly-N$_2$H$_2$Ph was prepared in 65% yield by the tryptic condensation of Boc-Asp(OBzl)-Arg-OMe and H-Phe-Gly-N$_2$H$_2$Ph (Figure 10). The above esterified acyl-group donor was available from the incubation of Boc-Asp(OBzl)-OMe and H-Arg-N$_2$H$_2$Ph in the presence of CPD-Y, following which the resulting dipeptide Boc-Asp(OBzl)-Arg-N$_2$H$_2$Ph (yield: 48%) was rearranged to yield the corresponding dipeptide methyl ester in 51% yield. All attempts to prepare Boc-Asp(OBzl)-Arg-N$_2$H$_2$Ph from Boc-Asp(OBzl)-OMe or Boc-Asp(OBzl)-OH via chymotrypsin- and thermolysin catalysis, respectively, had failed. Furthermore, the use of aspartic acid derivatives having a free β-carboxylate function did not result in the formation of the desired dipeptide.

In addition to the above-mentioned MSH fragments, another tetrapeptide adjacent to the N-terminus of the "active core" of dogfish β-MSH[49] was synthesized. The synthetic pathway leading to the tetrapeptide Boc-Tyr-Lys-Phe-Gly-N$_2$H$_2$Ph (Figure 11) was suggested by the presence of two aromatic residues resulting in putative chymotrypsin-sensitive bonds in the primary structure. This feature was exploited for the chymotryptic synthesis of the dipeptides Boc-Tyr-Lys-N$_2$H$_2$Ph (56%) and Boc-Phe-Gly-N$_2$H$_2$Ph (68%) (Figure 11). After rearranging Boc-Tyr-Lys-N$_2$H$_2$Ph to the corresponding dipeptide methyl ester, the latter was reacted with H-Phe-Gly-N$_2$H$_2$Ph in the presence of trypsin to give Boc-Tyr-Lys-Phe-Gly-N$_2$H$_2$Ph in 60% yield.

A combination of enzymatic and chemical steps was used for the preparation of a protected eledoisin$_{6\text{-}11}$-hexapeptide (Figure 12).[50] However, in contrast to the previously mentioned syntheses of biological peptides or peptide fragments, the authors performed the enzymatic reactions in aqueous-organic biphasic systems. α-Chymotrypsin- and papain-catalyzed syntheses were enhanced both by the use of "preactivated" esterified acyl-group donors and by the precipitation of the resulting products. Furthermore, the extent of synthesis was markedly improved by using a twofold molar excess of the respective amine component.

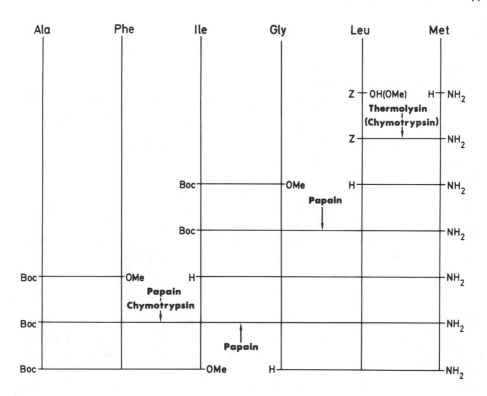

FIGURE 12.   Synthesis of the eledoisin$_{6-11}$ fragment. (The unlabeled peptide bonds were prepared by conventional, chemical methods).

The biphasic systems which were composed of either a papain- or a chymotrypsin-specific buffer and carbon tetrachloride were chosen in order to favor the precipitation of the prospective products. Other work showed that the reaction yields were drastically diminished when a water-immiscible trichloroethylene/petroleum ether mixture or the water-miscible methanol were used in place of carbon tetrachloride. Yet, the degree of synthesis was not solely dependent on the chemical nature of the organic layer but was also affected by the volume ratio of the two phases. In the presence of chymotrypsin the hexapeptide Boc-Ala-Phe-Ile-Gly-Leu-Met-NH$_2$ was obtained in maximum yield from its constituent di- and tetrapeptide (cf. Figure 12) only when carbon tetrachloride formed 60% by volume of the solvent system. Any increase or decrease in this figure resulted in reduced yields. Starting from the same educts, the eledoisin hexapeptide could also be obtained via papain catalysis (Figure 12). Unlike the chymotryptic synthesis, the optimal volume ratio of organic to aqueous phase amounted to 40 to 50% in the presence of papain. An alternative, synthetic route to the eledoisin hexapeptide involved the papain-catalyzed condensation of two tripeptide fragments (cf. Figure 12). In this case, however, the reaction yields remained unchanged when the percentage of organic co-solvent (CCl$_4$) was increased from 10 to 50%.

The above findings are compatible with theoretical studies on enzymatic syntheses in aqueous-organic biphasic systems.[51] With water as one of the reaction products, the yield of the other product, in the present case a peptide derivative, is a rather complex function of the ratio of the organic and aqueous phase (cf. Chapter 5, Equation 16). Depending upon the partition coefficients of a given set of reactants, i.e., the degree to which they are divided between two solvents, the volume ratio dependence of the peptide product can, for instance, pass through a maximum (cf. Chapter 5, Figure 3). With a different set of reactants displaying an altered partition behavior, variation of the volume ratio of the two phases within a certain range will have a negligible effect on the product yield (cf. Chapter 5, Figure 3). Consequently, the difference in the optimum phase volume ratios observed during the chymo-trypsin- and papain-catalyzed synthesis may be ascribed to the protease-specific buffers, the distinct chemical composition of which may lead to different partition coefficients of the very same reactants. Furthermore, while the harmful effects of organic co-solvents on enzyme stability can be greatly reduced in biphasic systems, they may not be completely eliminated.[52] Thus in the foregoing example, it may be that papain is more sensitive than chymotrypsin to the effects of the organic solvent carbon tetrachloride.

Although the previous examples demonstrate the usefulness of the "biphasic" technique, it may nevertheless prove a difficult matter to select the optimal conditions for enzymatic syntheses in biphasic systems.

Recently, a further mixed chemo-enzymic approach to the synthesis of partially protected fragments of the mouse epidermal growth factor (EGF) has been reported.[53] EGF is a single-chain polypeptide containing 53 amino acid residues with three disulfide bridges (between residues 6 and 20, 14 and 31, 33, and 42) which define three loop-regions including the residues 1 to 20, 14 to 31, and 32 to 53.[54,55] The intact EGF peptide plays a prominent role in the regulation of cell proliferation.[56] However, some subunits of the hormone are also capable of eliciting EGF-like activities.[57] This study described the preparation of three fragments corresponding to the positions (3 to 14), (21 to 31), and (33 to 42) of the primary structure of EGF which were projected to be precursors for the synthesis of the entire hormone. The proteases used as catalysts for the various peptide-bond-forming steps were carboxypeptidase Y, trypsin, chymotrypsin, papain, chymopapain, elastase, and protease V8 (endoproteinase Glu-C) from *Staphylococcus aureus,* strain V8. Esterified acyl-group donors were used as carboxyl components during the protease-mediated coupling steps so that peptide bond formation was kinetically controlled. Protease-labile α-amino protecting groups were introduced in several instances to influence the hydrophobicity or the hydrophilicity of the respective reactants. The protected undecapeptide Bz-[Gly$_{21}$Cys$_{31}$(Acm)]EGF(21-31)NHEt was finally assembled by the condensation of the three building blocks which had been separately prepared (Figure 13). The protected subunit EGF[21-24] was synthesized via trypsin-catalyzed coupling of the dipeptides Bz-Gly-His-OMe and H-Ile-Glu-(OBzl)-OMe which had been prepared chemically in the presence of DCC and HOSu. The protected subunit EGF$_{25-27}$ was assembled as tetrapeptide because Bz-Arg-OEt was introduced as N-terminal protecting group via trypsin-catalyzed reaction with H-Ser-OEt.

This dipeptide was C-terminally elongated by H-Leu-OH in the presence of carboxypep-tidase Y. The resulting tripeptide Bz-Arg-Ser-Leu-OH was then reacted with H-Asp(OBzl)-OH in the presence of the racemization-inducing, activating agent DCC to give the desired EGF$_{25-27}$ subunit. The protected EGF$_{28-31}$ segment — the α-amino protector group was again provided by the Bz-Arg moiety — was prepared by fragment condensation as outlined in Figure 13; thus after introduction of the Bz-Arg-protection via trypsin-catalysis, the resulting dipeptide Bz-Arg-Ser-OEt was coupled to H-Tyr-NH$_2$ in the presence of carboxypeptidase Y to give the tripeptide Bz-Arg-Ser-Tyr-NH$_2$. The fragment Bz-Phe-Thr-Cys(Acm)-NHEt, in which the Boc-Phe moiety served as a chymotryptically introduceable α-amino protection, was synthesized from Bz-Phe-Thr-OEt and H-Cys(Acm)-NHEt in the presence of chymo-papain. As chymopapain elicits a papain-like *(vide supra,* Chapter 7, Section VIII) primary specificity for aromatic amino acids in the S$_2$-site of the substrates, the phenylalanine residue served not only as protecting group but also constituted an appropriate S$_2$-site. After chy-motryptic removal of the Bz-Phe protection the resulting dipeptide was coupled to the tripeptide Bz-Arg-Ser-Tyr-OMe in the presence of chymotrypsin to give the targeted pen-tapeptide. (The tripeptide methylester had been prepared by chymopapain-catalyzed deam-idation of the corresponding tripeptide amide *[vide supra]* followed by the actual acid-catalyzed methylation step.) The next step involved the removal of the Boc-Arg protection

79

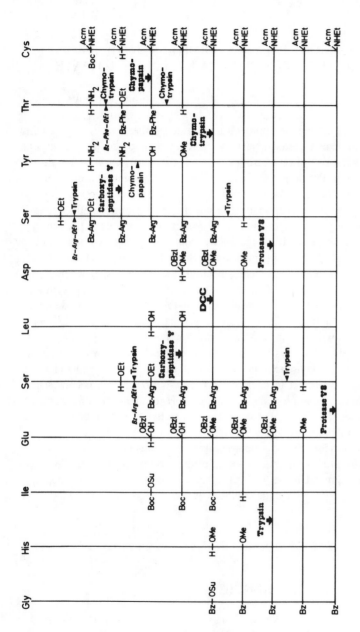

FIGURE 13. Enzymo-chemical synthesis of the protected EGF$_{21-31}$ fragment.

from the pentapeptide and the resulting tetrapeptide was subsequently coupled to the tetra-peptide Bz-Arg-Ser-Leu-Asp-OMe in the presence of protease V8. (The primary specificity of this endoprotease is directed at aspartate and glutamate residues occupying the S_1-site of the respective substrates.)[58] The resulting octapeptide was deprotected, i.e., the N-terminal Bz-Arg grouping was tryptically removed, and the heptapeptide fragment EGF_{25-31} was reacted with the tetrapeptide fragment EGF_{21-24} via protease V8 catalysis to yield the final protected octapeptide EGF_{21-31}.

V. PROTEASES AS BIOCATALYSTS FOR THE SYNTHESIS OF BIOLOGICALLY ACTIVE PEPTIDES — A SUMMARY TO DATE

The successful preparative synthesis, via protease catalysis, of biologically active analogs of naturally occurring peptides as described in the preceding chapters, provided convincing proof of the capabilities of the enzymatic approach to peptide synthetic chemistry. Owing to the property of protease-immanent stereospecificity, the enzymatic procedure provides an outstanding way of avoiding the risk of racemization which is frequently encountered during chemical fragment condensation. Furthermore, the regiospecific action of proteases facilitates the preparation of complex peptides, whose synthesis by chemical means is often impeded by sequence-dependent side reactions. Despite these promising features, enzymatic synthesis has not yet reached the level of general applicability. The main problem lies in the shortage of a complete set of proteases whose stringent specificity would enable the synthetic peptide chemist to selectively prepare all possible peptide bonds. Since this requirement is unlikely to be satisfied by nature, the inevitable incomplete collection of stringently specific proteases must be supplemented by proteases whose broader substrate specificity enables the synthesis of a variety of peptide bonds. However, the bright side of this picture may be darkened since the lack of narrowly restricted specificity would jeopardize preexisting peptide bonds. Nevertheless, it has been demonstrated in several of the previously mentioned enzymatic syntheses that putative synthetic pathways which prove to be dead ends can be by-passed by judicious selection of alternate synthetic routes. In case an expedient is not at hand, there always remains the possibility to resort to chemical procedures. When this becomes nec-essary, it is preferable to have a glycine residue at the COOH-terminal end of a carboxyl component to be coupled since the presence of the optically inactive glycine moiety, though representing a "critical" amino acid residue in the light of enzymatic syntheses, avoids any hazard to the chiral integrity of the product. In specific cases proline residues may be similarly exploited. Thus while the full proteosynthetic potential of the proline specific proteases has yet to be documented, the chemical route to the synthesis of proline-containing peptides offers an acceptable alternative to enzymatic methods given the greatly reduced risk of product racemization.

REFERENCES

1. **Hughes, J., Smith, T. W., Kosterlitz, H. W., Fothergill, L. A., Morgan, B. A., and Harris, H. R.,** Identification of two related pentapeptides from the brain with potent opiate agonist activity, *Nature (London)*, 258, 577, 1975.
2. **Kullmann, W.,** Enzymatic synthesis of Leu- and Met-enkephalin, *Biochem. Biophys. Res. Commun.*, 91, 693, 1979.
3. **Kullmann, W.,** Proteases as catalysts of enzymic syntheses of opioid peptides, *J. Biol. Chem.*, 255, 8234, 1980.
4. **Milne, H. B. and Carpenter, F. H.,** Peptide synthesis via oxidation of *N*-acyl-α-amino acid phenylhy-drazides. III. Dialanyl-insulin and diphenylalanyl-insulin, *J. Org. Chem.*, 33, 4476, 1968.

5. **Milne, H. B. and Kildey, W.,** Peptide synthesis via oxidation of *N*-acetyl-α-amino acid phenylhydrazides, *J. Org. Chem.,* 30, 64, 1965.

6. **Ramachandran, L. K. and Witkop, B.,** *N*-bromosuccinimide cleavage of peptides, *Methods Enzymol.,* 11, 283, 1967.

7. **Peterson, R. L., Hubele, K. W., and Niemann, C.,** The α-chymotrypsin-catalyzed hydrolysis of α-*N* and O-alkyl derivatives of α-N-acetyl-L-tyrosine methyl ester, *Biochemistry,* 2, 942, 1963.

8. **Zapevalova, N. P., Gorbunova, E. Y., and Mitin, Y. V.,** Synthesis of leucine- and methionine-enkephalin using papain (in Russian), *Bioorg. Khim.,* 11, 733, 1985.

9. **Mitin, Y. V., Zapevolova, N. P., and Gorbunova, E. Y.,** Peptide synthesis catalyzed by papain at alkaline pH values, *Int. J. Peptide Protein Res.,* 23, 528, 1984.

10. **Widmer, F., Breddam, K., and Johansen, J. T.,** Carboxypeptidase Y as a catalyst for peptide synthesis in aqueous phase with minimal protection, in *Peptides 1980, Proc. 16th Europ. Peptide Symp.,* Brunfeldt, K., Ed., Scriptor, Kopenhagen, 1981, 46.

11. **Widmer, F., Breddam, K., and Johansen, J. T.,** Carboxypeptidase Y catalyzed peptide synthesis using amino acid alkyl esters as amine components, *Carlsberg Res. Commun.,* 45, 453, 1980.

12. **Breddam, K., Widmer, F., and Johansen, J. T.,** Carboxypeptidase Y catalyzed transpeptidations and enzymatic peptide synthesis, *Carlsberg Res. Commun.,* 45, 237, 1980.

13. **Kullmann, W.,** Design, synthesis, and binding properties of an opiate receptor mimetic peptide, *J. Med. Chem.,* 27, 106, 1984.

14. **Schwartz, J.-C., Malfroy, B., and De La Baume, S.,** Biological inactivation of enkephalins and the role of enkephalin-dipeptidyl-carboxypeptidase ("Enkephalinase") as neuropeptidase, *Life Sci.,* 29, 1715, 1981.

15. **Di Maio, J., Nguyen, T., M.-D., Lemieux, C., and Schiller, P. W.,** Synthesis and pharmacological characterization in vitro of cyclic enkephalin analogues: effect of conformational constraints on opiate receptor selectivity, *J. Med. Chem.,* 25, 1432, 1982.

16. **Paterson, S. J., Robson, L. E., and Kosterlitz, H. W.,** Opioid receptors, in *The Peptides,* Vol. 6, Gross, E., Meienhofer, J., and Hruby, V., Eds., Academic Press, New York, 1984, 147.

17. **Stoineva, I. B. and Petkov, D. D.,** Chemical-enzymatic incorporation of D-amino acids into peptides: synthesis of diastereomeric (D-Ala$_2$, D-Leu$_5$) enkephalinamides, *FEBS Lett.,* 183, 103, 1985.

18. **Petkov, D. D. and Stoineva, I. B.,** Enzyme peptide synthesis by an iterative procedure in a nucleophile pool, *Tetrahedron Lett.,* 25, 3751, 1984.

19. **Morihara, K. and Oka, T.,** α-Chymotrypsin as the catalyst for peptide synthesis, *Biochem. J.,* 163, 531, 1977.

20. **Kullmann, W.,** unpublished results.

21. **Fruton, J. S.,** Proteinase-catalyzed synthesis of peptide bonds, *Adv. Enzymol. Relat. Areas Mol. Biol.,* 53, 239, 1982.

22. **Kullmann, W.,** Enzymatic synthesis of dynorphin(1-8), *J. Org. Chem.,* 47, 5300, 1982.

23. **Minamino, N., Kangawa, K., Fukuda, A., Matsuo, H., and Jagarashi, H.,** A new opioid octapeptide related to dynorphin from porcine hypothalamus, *Biochem. Biophys. Res. Commun.,* 95, 1475, 1980.

24. **Seizinger, B. R., Höllt, V., and Hertz, A.,** Evidence for the occurrence of the opioid octapeptide dynorphin (1-8) in the neurointermediate pituitary of rats, *Biochem. Biophys. Res. Commun.,* 102, 197, 1981.

25. **Goldstein, A., Tachibana, S., Lowney, L. I., Hunkapillar, M., and Hood, L.,** Dynorphin(1-13), an extraordinary potent opioid peptide, *Proc. Natl. Acad. Sci. U.S.A.,* 76, 6666, 1979.

26. **Tachibana, S., Araki, K., Ohya, S., and Yoshida, S.,** Isolation and structure of dynorphin, an opioid peptide from porcine duodenum, *Nature (London),* 295, 339, 1982.

27. **Chavkin, C. and Goldstein, A.,** Specific receptor for the opioid peptide dynorphin: structure-activity relationships, *Proc. Natl. Acad. Sci. U.S.A.,* 78, 6543, 1981.

28. **Wünsch, E.,** Synthese von Peptiden, in *Houben-Weyl, Methoden der organischen Synthese,* Vol. 15, Part 1, Thieme Verlag, Stuttgart, 1974.

29. **Bodanszky, M. and Martinek, J.,** Side-reactions in peptide synthesis, *Synthesis,* 333, 1981.

30. **Geiger, R. and König, W.,** Amine protecting groups, in *The Peptides,* Vol. 3, Gross, E. and Meienhofer, J., Eds., Academic Press, New York, 1981, 1.

31. **Barany, G. and Merrifield, R. B.,** Solid phase synthesis, in *The Peptides,* Vol. 2, Gross, E. and Meienhofer, J., Eds., Academic Press, New York, 1980, 1.

32. **Kangawa, K., Minamino, N., Chino, N., Sakakibara, S., and Matsuo, H.,** The complete amino acid sequence of α-neo-endorphin, *Biochem. Biophys. Res. Commun.,* 99, 871, 1981.

33. **Mutt, V. and Jorpes, J. E.,** Structure of porcine cholecystokinin-pancreozymin, *Eur. J. Biochem.,* 6, 156, 1968.

34. **Kullmann, W.,** Protease-catalyzed peptide bond formation: application to synthesis of the COOH-terminal octapeptide of cholecystokinin, *Proc. Natl. Acad. Sci. U.S.A.,* 79, 2840, 1982.

35. **Anastasi, A., Erspamer, V., and Endean, R.,** Isolation and structure of caerulein, an active decapeptide from the skin of Hyla caerulea, *Experientia,* 23, 699, 1967.

36. **Erspamer, V., Bertaccimi, G., de Caro, G., Endean, R., and Impicciatore, M.,** Pharmacological actions of caerulein, *Experientia,* 23, 702, 1967.

37. **Takai, H., Sakato, K., Nakamizo, N., and Isowa, Y.,** Enzymatic synthesis of caerulein peptide, in *Peptide Chemistry 1980,* Okawa, K., Ed., Protein Research Foundation, Osaka, Japan, 1981, 213.

38. **Morihara, K.,** Comparative specificity of microbial proteinases, *Adv. Enzymol. Relat. Areas Mol. Biol.,* 41, 179, 1974.

39. **Iio, I. and Yamasaki, M.,** Specificity of acid proteinase A from *Aspergillus niger* var. macrosporus towards B-chain of performic acid oxidized insulin, *Biochim. Biophys. Acta,* 429, 912, 1976.

40. **Kullmann, W.,** Protease-catalyzed synthesis of melanocyte-stimulating hormone (MSH) fragments, *J. Protein Chem.,* 2, 289, 1983.

41. **Yajima, H. and Kawatami, H.,** Synthesis of ACTH-active peptides and analogs, in *Chemistry and Biochemistry of Amino Acids, Peptides, and Proteins,* Vol. 2, Weinstein, B., Ed., Marcel Dekker, New York, 1974, 39.

42. **Dayhoff, M. O.,** *Atlas of Protein Sequence and Structure,* Vol. 5, National Biomedical Research Foundation, Washington, D.C., 1972, D-194.

43. **Otsuka, H. and Inouye, K.,** Syntheses of peptides related to the N-terminal structure of corticotropin. III. The synthesis of L-histidyl-L-phenylalanyl-L-arginyl-L-tryptophan, the smallest peptide exhibiting the melanocyte-stimulating and the lipolytic activities, *Bull. Chem. Soc. Jpn.,* 37, 1465, 1964.

44. **Hofmann, K., Woolner, M. E., Spühler, G., and Schwartz, E. T.,** Studies on polypeptides. X. The synthesis of a pentapeptide corresponding to an amino acid sequence present in corticotropin and in the melanocyte stimulating hormones, *J. Am. Chem. Soc.,* 80, 1486, 1958.

45. **Schwyzer, R. and Li, C. H.,** A new synthesis of the pentapeptide L-histidyl-L-phenylalanyl-L-arginyl-L-tryptophyl-glycine and its melanocyte-stimulating activity, *Nature (London),* 182, 1669, 1958.

46. **Tauber, H.,** Phenylalanylphenylalanine ethyl ester synthesis, *J. Am. Chem. Soc.,* 74, 847, 1952.

47. **Anderson, G. and Luisi, P. L.,** Papain-induced oligomerization of amino acid esters, *Helv. Chim. Acta,* 62, 488, 1979.

48. **Isowa, Y. and Ichikawa, T.,** Syntheses of *N*-acyl dipeptide derivatives by metalloproteinases, *Bull. Chem. Soc. Jpn.,* 52, 796, 1979.

49. **Bennett, H. P. L., Lowry, P. J., McMartin, C., and Scott, A. P.,** Structural studies of α-melanocyte-stimulating hormone and a novel β-melanocyte-stimulating hormone from the neurointermediate lobe of the pituitary of the dogfish squalus acanthias, *Biochem. J.,* 141, 439, 1974.

50. **Kuhl, P., Döring, G., Neubert, K., and Jakubke, H.-D.,** Synthese von N-geschütztem Eledoisin(6-11)-Hexapeptid unter Verwendung von Proteasen als Biokatalysatoren, *Monatsh. Chem.,* 115, 423, 1984.

51. **Martinek, K., Semenov, A. N., and Berezin, I. V.,** Enzymatic synthesis in biphasic aqueous-organic systems. I. Chemical equilibrium shift, *Biochim. Biophys. Acta,* 658, 76, 1981.

52. **Kuhl, P., Walpuski, J., and Jakubke, H.-D.,** Untersuchungen zum Einfluss der Reaktionsbedingungen auf die α-chymotrypsinkatalysierte Peptidsynthese im wässrig-organischen Zweiphasensystem, *Pharmazie,* 37, 766, 1982.

53. **Widmer, F., Bayne, S., Houen, G., Moss, B. A., Rigby, R. D., Whittaker, R. G., and Johansen, J. T.,** Use of proteolytic enzymes for synthesis of fragments of mouse epidermal growth factor, in *Peptides 84, Proc. 18th Eur. Peptide Symp.,* Ragnarsson, U., Ed., Almquist and Wiksell, Stockholm, 1984, 193.

54. **Savage, C. R., Jr., Inagami, T., and Cohen, S.,** The primary structure of epidermal growth factor, *J. Biol. Chem.,* 247, 7612, 1972.

55. **Savage, C. R., Jr., Hash, J. H., and Cohen, S.,** Epidermic growth factor. Location of disulfide bonds, *J. Biol. Chem.,* 248, 7669, 1973.

56. **Gregory, H.,** Isolation and structure of urogastrone and its relationship to epidermal growth factor, *Nature (London),* 257, 325, 1975.

57. **Komoriya, A., Hortsch, M., Meyers, C., Smith, M., Kanety, H., and Sclesinger, J.,** Biologically active synthetic fragments of epidermal growth factor: localization of a major receptor binding region, *Proc. Natl. Acad. Sci. U.S.A.,* 81, 1351, 1984.

58. **Drapeau, G. R., Boily, Y., and Houmard, J.,** Purification and properties of an extracellular protease of *Staphylococcus aureus, J. Biol. Chem.,* 247, 6720, 1972.

Chapter 9

PROTEASE-CATALYZED OLIGOMERIZATION

During the early decades of the present century, the phenomenon of so-called plastein formation attracted particular attention because it appeared to support the concept of proteinbiosynthesis by reversible proteolysis *(vide supra,* Chapter 2). The chemical nature of the plasteins, which are usually formed as precipitates upon addition of proteases to concentrated peptic digests, was subject to considerable discussion for a long time.[1] Systematic studies of Virtanen,[2] and Wieland, Determann, and co-workers, on the pepsin-induced assembly of synthetic peptides,[3-6] revealed that plasteins are products of protease-controlled oligomerization of suitable plastein-active peptide monomers. The first report dealing with the isolation and characterization, from a proteolytic partial hydrolysate of homogeneous peptide substrates (an octa- and an undecapeptide) for the pepsin-catalyzed formation of plasteins was published in 1960.[3] In 1961 Determann and Wieland[4] described the synthesis of an artificial plastein-active pentapeptides, the amino acid sequence of which read: H-Tyr-Ile-Gly-Glu-Phe-OH. Upon pepsin-treatment at pH 5 this peptide was condensed to give an insoluble oligomer with an average degree of polymerization of 2.5. In respect of the molecular mechanism of the plastein formation was concerned, the authors showed that the assembly of the oligomer represented a condensation reaction rather than a transpeptidation. Further studies indicated that in the presence of pepsin the plastein-activity of the peptide monomers largely depended on two requirements. First, the peptide chain must be composed of at least four amino acid units;[5] and second, the C-terminal position of the peptide monomer must be occupied by a phenylalanine-, a tyrosine-, or a leucine residue of L-configuration.[6] The average polymerization degree of all the pepsin-induced plastein syntheses studied to date ranges from 2.3 to 11.[7]

"Modified" plastein reactions have found practical applications in the chemistry of foods where they have been used to improve the nutritional quality of food proteins by incorporation of essential amino acids.[8] In this case the combined actions of enzymatic proteolysis and subsequent papain-mediated proteosynthesis in the presence of various amino acid ethyl esters resulted in plasteins enriched with the desired amino acid derivative.

Protease-catalyzed oligomerization of natural or synthetic peptide monomers, not possessing classic plastein forming activity, was reported by Fruton and co-workers.[9-12] These authors observed that cathepsin C, a thiol protease also known as "dipeptidyl aminotransferase", which possesses a chymotrypsin-like specificity, catalyzed the oligomerization of the following dipeptide amides in the pH-range of 6.6 to 7.8: H-Gly-Phe-NH$_2$, H-Gly-Tyr-NH$_2$, H-Gly-Trp-NH$_2$, H-Ala-Phe-NH$_2$, H-Ala-Tyr-NH$_2$, H-Gly-Tyr(CH$_3$)-NH$_2$, H-Lys(ϵ-Ac)-Phe-NH$_2$. The sparingly soluble products were present mainly as octa- or decapeptide amides. However, oligomers were not formed either when amino acid amides, tri- or tetrapeptide amides were used as reactants, or when dipeptidyl amides were incubated at pH 5.2, under which conditions the proteolytic activity of cathepsin C is optimal. Successful chain propagation proceeded by stepwise addition, via consecutive transamidation reactions, of dipeptidyl units to the carboxyl end of the growing chain.[13]

A more recent study has described the isolation from *Streptomyces cellulosae* of a protease which catalyzes the oligomerization of free dipeptides to yield tetra- and hexapeptides.[14] The following oligomeric peptides could be obtained: (Leu-Gly)$_3$ and (Leu-Gly)$_2$ from Leu-Gly, (Val-Phe)$_3$ and (Val-Phe)$_2$ from Val-Phe, (Leu-Met)$_3$ and (Leu-Met)$_2$ from Leu-Met, and (Phe-Val)$_2$ from Phe-Val.

The trypsin-mediated oligomerization of the tripeptide H-Ala-Ala-Arg-OH was described by Čeřovský and Jošt.[15] Maximal yields were obtained in the pH range between 6.2 and 6.5 and could not be improved by the addition of organic solvents. The product of the enzymatic process was identified as a hexamer. In contrast, the incubation of H-Ala-Arg-OH, H-Arg-OH, or H-Arg-OMe in the presence of trypsin, did not yield any oligomeric products. Presumably, a tripeptide represents the minimal chain-length requirement for a successful oligomerization process.

The foregoing studies on trypsin-induced oligomer formation have shown that peptides of a certain chain-length are suitable substrates for this process but amino acid derivatives are not. However, the following examples demonstrate that, given the right conditions, amino acid derivatives can also be enzymatically oligomerized. Thus, α-chymotrypsin-controlled oligomerization of several amino acid esters has been reported by Brenner and co-workers.[16-18] The tendency to oligomerize of the amino acid isopropylesters decreased in the following order: threonine, methionine, phenylalanine, and tyrosine; whereas tryptophan- and valine esters were merely esterolyzed. A variety of differently shaped esters was assayed but their chemical nature did not significantly influence the outcome of the enzymatic reactions. Optimal results were obtained at pH 9. The major oligomerization products were free di- and tripeptides, whereas peptide esters were formed to a minor extent.

The papain-catalyzed formation of oligopeptide methyl esters starting from leucine-, methionine-, phenylalanine-, tyrosine-, and tryptophan methyl esters was described by Luisi and co-workers.[19,20] Whereas H-Leu-OMe and H-Phe-OMe did not give oligomers unless a "starter-molecule" or "primer" such as Z-Gly-OH, Boc-Leu-OH or Boc-Gly-OH, was present in the incubation system, H-Tyr-OMe, H-Trp-OMe, and H-Met-OMe could be oligomerized in the absence of a primer. In contrast, the methyl esters of histidine, serine, and alanine, and in addition, the amides of leucine and phenylalanine, did not polymerize even in the presence of a primer. Maximal product yields during the polymerization of H-Leu-OMe with Z-Gly-OH present as a "primer" were achieved at a pH value of 5.6. Furthermore, oligomerization was crucially dependent on the nature and the concentration of the buffer used. Peptide bond synthesis proceeded only in a citrate buffer and maximal yields — generally amounting to 51 to 95% — and the most homogeneous products could only be obtained using the most concentrated buffer. The degree of polymerization, which ranged from 3 to 11, could not be increased by addition of ethanol. This was added with the idea of improving the solubility and thus the reactivity of the precipitating products but, on the contrary, the presence of ethanol decreased the reaction yields.

Carboxypeptidase Y catalyzed oligomerization was observed by Widmer et al.[21] Systematic studies at pH 9.6 using Boc-Ala-OMe as acyl group donor and various amino acid alkyl esters as nucleophilic acceptors revealed that the extent of oligomerization was very sensitive both to the nature of the side chain and the chemical nature of the alkyl ester. Whereas methyl esters of hydrophobic amino acids underwent oligomerization to varying degrees, the methyl esters of hydrophilic amino acids showed little or no tendency to oligomerize. In general, the extent of oligomerization decreased with increasing size of the alkyl ester group. The only exception was glycine, the ethyl ester of which was dimerized whereas the methyl ester did not give any oligomeric product. In fact this tendency of the glycine ethyl ester to dimerize was usefully exploited by Widmer et al. during the synthesis of Met-enkephalin.[22] Nevertheless, oligomerization processes usually represent undesired side reactions in the course of enzymatic syntheses and as shown by Breddam et al.,[23] these effects can be drastically reduced by modifying carboxypeptidase Y with phenylmercuric chloride.

REFERENCES

1. **Florkin, M. and Stotz, E. H.**, Reversible zymo-hydrolysis, in *Comprehensive Biochemistry*, Vol. 32, Elsevier, Amsterdam, 1977, 307.
2. **Virtanen, A. I.**, Über die enzymatische Polypeptidsynthese, *Makromol. Chem.*, 6, 94, 1951.
3. **Wieland, T., Determann, H., and Albrecht, E.**, Untersuchungen über die Plastein-Reaktion. Isolierung einheitlicher Plastein-Bausteine, *Justus Liebigs Ann. Chem.*, 633, 185, 1960.
4. **Determann, H. and Wieland, T.**, Ein synthetisches Pentapeptid als Plastein-Monomeres. Untersuchungen über die Plastein-Reaktion. II, *Makromol. Chem.*, 44-46, 312, 1961.
5. **Determann, H., Bonhard, K., Köhler, R., and Wieland, T.**, Untersuchungen über die Plastein-Reaktion. VI. Einfluss der Kettenlänge und der Endgruppen des Monomeren auf die Kondensierbarkeit, *Helv. Chim. Acta*, 46, 2498, 1963.
6. **Determann, H., Heuer, J., and Jaworek, D.**, Untersuchungen über die Plastein-Reaktion. VIII. Spezifität des Pepsins bei der Kondensationsreaktion, *Justus Liebigs Ann. Chem.*, 690, 189, 1965.
7. **Determann, H., Eggenschwiller, S., and Michel, W.**, Untersuchungen über die Plastein-Reaktion. VII. Molekulargewichtsverteilung des enzymatischen Kondensationsprodukts, *Justus Liebigs Ann. Chem.*, 690, 182, 1965.
8. **Fujimaki, M., Arai, S., and Yamashita, M.**, Enzymatic protein degradation and resynthesis for protein improvement, *Adv. Chem. Series*, 160, 156, 1977.
9. **Jones, M. E., Hearn, W. R., Fried, M., and Fruton, J. S.**, Transamidation reactions catalyzed by cathepsin C, *J. Biol. Chem.*, 195, 645, 1952.
10. **Fruton, J. S., Hearn, W. R., Ingram, V. M., Wiggans, D. S., and Winitz, M.**, Synthesis of polymeric peptides in proteinase-catalyzed transamidation reactions, *J. Biol. Chem.*, 204, 891, 1953.
11. **Izymiya, N., and Fruton, J. S.**, Specificity of cathepsin C, *J. Biol. Chem.*, 218, 59, 1956.
12. **Würz, H., Tanaka, A., and Fruton, J. S.**, Polymerization of dipeptide amides by cathepsin C, *Biochemistry*, 1, 19, 1962.
13. **Fruton, J. S.**, Proteinase-catalyzed synthesis of peptide bonds, *Adv. Enzymol. Relat. Areas Mol. Biol.*, 53, 239, 1982.
14. **Muro, T., Tominaga, Y., and Okada, S.**, Formations of oligopeptides by the protease from *Streptomyces cellulosae*, *Agric. Biol. Chem.*, 48, 1231, 1984.
15. **Čeřovský, V., and Jošt, K.**, Trypsin-catalyzed oligomerization of the Ala-Ala-Arg tripeptide, in *Peptides 1982, 17th Eur. Peptide Symp.*, Blahá, K. and Maloň, P., Eds., Walter de Gruyter, Berlin, 1983, 395.
16. **Brenner, M., Müller, H. R., and Pfister, R. W.**, Eine neue enzymatische Peptidsynthese, *Helv. Chim. Acta*, 33, 568, 1950.
17. **Brenner, M. and Pfister, R. W.**, Enzymatische Peptidsynthese. Isolierung von enzymatisch gebildetem L-Methionyl-L-methionin und L-Methionyl-L-methionyl-L-methionin; Vergleich mit synthetischen Produkten, *Helv. Chim. Acta*, 34, 2085, 1951.
18. **Brenner, M., Sailer, E., and Rüfenacht, K.**, Enzymatische Peptidsynthese. Peptidbildung aus DL-Threonin-isopropylester, *Helv. Chim. Acta*, 34, 2096, 1951.
19. **Anderson, G. and Luisi, P. L.**, Papain-induced oligomerization of α-amino acid esters, *Helv. Chim. Acta*, 62, 488, 1979.
20. **Jost, R., Brambilla, E., Monti, J. C., and Luisi, P. L.**, Papain-catalyzed oligomerization of α-amino acids. Synthesis and characterization of water-insoluble oligomers of L-methionine, *Helv. Chim. Acta*, 63, 375, 1980.
21. **Widmer, F., Breddam, K., and Johansen, J. T.**, Carboxypeptidase Y catalyzed peptide synthesis using amino acid alkyl esters as amine components, *Carlsberg Res. Commun.*, 45, 453, 1980.
22. **Widmer, F., Breddam, K., and Johansen, J. T.**, Carboxypeptidase Y as catalyst for peptide synthesis in aqueous phase with minimal protection, in *Peptides 1980, Proc. 16th Eur. Peptide Symp.*, Brunfeldt, K., Ed., Scriptor, Kopenhagen, 1981, 46.
23. **Breddam, K., Widmer, F., and Johansen, J. T.**, Amino acid methyl esters as amine components in CPD-Y catalyzed peptide synthesis: control of side-reactions, *Carlsberg Res. Commun.*, 48, 231, 1983.

Chapter 10

PROTEASE-CATALYZED SEMISYNTHESES

I. INTRODUCTION

In contrast to the total synthesis of peptides, so-called "semisynthesis" delivers peptides or proteins which are composed both of "natural" and "artificial" constituents. This approach involves the replacement of a peptide fragment or a single amino acid of the native peptide chain by a structurally altered, synthetic fragment or a different amino acid. (The designation "semisynthesis" appears rather euphemistic, because the portion which is actually altered usually constitutes far less than 50% of the original protein.) As a result, "mutated" peptides or proteins are generated, the structural and functional properties of which may differ from those of the parent molecule. These modified analogs of native peptides or proteins can represent valuable tools for biochemical, molecular, biological, or medicinal investigations.

Evidently, a complete *de novo* synthesis would offer an enhanced versatility in designing modifications, but the preparation of large-sized peptides or proteins either is a very intricate task or is beyond the scope of current peptide synthetic technology. Considering the present efficiency of peptide synthesis one must still agree with the judgment of Gross and Meienhofer who stated (citation): "It would seem then, that at this time either solution synthesis or solid phase synthesis, or combinations of both, may be utilized to synthesize peptides of up to 15 residues readily. Those of up to 30 or 40 residues will require considerable expertise. For peptides of a size approaching that of proteins, the expenditure of great efforts and patience will be essential."[1] Semisynthesis, therefore, offers a strategy to avoid these difficulties. However, despite its promise, the semisynthetic approach is not without its problems. Thus, the unspecific action of activating agents, which are required for the "chemical" coupling of natural and synthetic intermediates, can affect the integrity of both chirality and the side-chain functionalities. Although protecting groups are employed to suppress the reactivity of side-chain functions, the introduction and removal of these protector groups requires additional effort and may give rise to side reactions. However, these drawbacks can be largely eliminated if a protease can be used to stereo- and regiospecifically catalyze the coupling step. The specificity of the enzyme ensures the maintenance of chiral integrity and renders unnecessary the protection of functional groups. These advantages account for the growing number of enzymatic semisyntheses, some of which will be reviewed in the following chapter.

II. INSULIN

As millions of diabetics suffer from an insulin deficiency, there exists a strong demand for human insulin. For evident reasons, this requirement cannot be satisfied by exploiting natural sources. On the other hand, the application of insulins of nonhuman origin may cause serious problems during long-lasting therapies. Therefore, it does not come as a surprise that numerous semisynthetic studies have focused on the preparation of human insulin. The pancreatic hormone consists of two peptide chains (A- and B-chain) which are cross-linked by two disulfide bonds[2] (cf. Figure 1). These two interchain and additional intrachain disulfide bridges are mainly responsible for the difficulties often encountered during total syntheses of insulin. In spite of promising developments, the *ab initio* synthesis of insulin remains an exceptionally intricate task. For these reasons and in view of the finding that the primary structures of human and porcine insulin differ only in the C-terminal amino acid

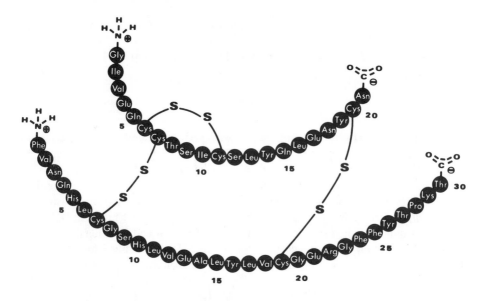

FIGURE 1.   Primary structure of human insulin.[2]

residue of the B-chain — threonine (cf. Figure 1) in place of alanine — a semisynthetic approach offers a realistic strategy to the preparation of human insulin.

The first protease-mediated preparation of partially synthetic human insulin was described by Inouye et al.[3] A desoctapeptide (B_{23}-B_{30})-insulin (DOI) obtained via tryptic cleavage of porcine insulin was N-acylated to provide $N^{\alpha A1}$, $N^{\alpha B1}$-(Boc)$_2$-DOI. The acylated compound could then serve as a carboxyl component in a prospective trypsin-controlled coupling step, since the C-terminal position of the truncated B-chain (B_{22}) was occupied by an arginine residue. The amine component, a synthetic octapeptide, ($N^{\epsilon B29}$-(Boc)(B_{23-30})-insulin — having, like human insulin, a threonine residue in position B_{30} — was prepared by solution methods. The side chain of Lys$_{B29}$ was acylated in order to confine the tryptic action to the formation of the B_{22}-B_{23} (Arg-Gly) bond. The *ex natura* carboxyl component and a tenfold molar excess of the *ex arte* amine component were incubated at pH 6.5 in the presence of trypsin. After 20 hr at 37°C the protected DOI had been converted to $N^{\alpha A1}$, $N^{\alpha N1}$, $N^{\alpha B29}$(Boc)$_3$-human insulin. The product yield came to 58%; so that the enzymatic approach worked more efficiently than comparable chemical techniques.[4,5]

Since the conversion of porcine into human insulin only requires the substitution of a C-terminal alanine unit of the B-chain by a threonine residue, the replacement of an octapeptide may appear overdone. However, a des-Ala$_{B30}$-insulin (DAI) derivative, represents the most suitable starting material for the build-up of semisynthetic human insulin. Unfortunately, the selective, tryptic cleavage of the Lys$_{B29}$-Ala$_{B30}$ bond is rather unlikely, because the Arg$_{B22}$-Gly$_{B23}$ linkage is concurrently split.[6,7] Nevertheless, due to the observation of Harris that carboxypeptidase A-catalyzed release of Ala$_{B30}$ is about eight times that of asparagine A$_{21}$,[8] Schmitt and Gattner developed a method which allowed the selective removal of Ala$_{B30}$ from porcine insulin.[9] Applying this technique Morihara et al.[10] obtained the truncated insulin derivate DAI, which was then incubated at pH 6.5 with a 50-fold molar excess of H-Thr-OBu^t in the presence of trypsin. After 20 hr at 37°C, 73% of DAI had been transformed into (Thr-OBu^t-B_{30})-insulin. A similar approach semisynthetic human insulin was described by Gattner et al.[11] who used H-Thr-OMe instead of H-Thr-OBu^t as the replacing agent. The observation of Morihara et al.[12] that trypsin did indeed catalyze the formation of the Lys$_{B29}$-Thr$_{B30}$ bond but in addition, although to a minor extent, the cleavage of the Arg$_{B22}$-Gly$_{B23}$ bond, is hardly surprising. The authors could suppress the undesired side reaction by introducing a protecting group for the guanidino function of Arg_{B22}. An alternative way to maintain the integrity of the trypsin-labile Arg_{B22}-Gly_{B23} bond was proposed by Jonczyk and Gattner.[13] The human insulin-Thr_{B30} *t*-butyl ester could be obtained via a "one-pot" reaction during which porcine insulin was incubated with H-Thr-OBu[t] in the presence of trypsin. It seems conceivable that in this case the conversion of porcine- into human insulin proceeded via the following steps. First, an intermediate DAI-trypsin complex is formed concurrently with the release of free alanine; following which the complex is instantaneously deacylated via H-Thr-OBu[t]-mediated aminolysis to give the targeted human insulin *tert*-butyl-ester, i.e., the overall reaction represents a tryptic transpeptidation process, and the DAI-trypsin-complex was not cleaved hydrolytically to release free DAI. The optimal product yield (70%) was achieved at pH 6.8 with the concentration of the amine component being 130 times greater than that of the carboxyl component while the ratio of enzyme- to substrate concentration was as much as 1:10. The desoctapeptide, $(B_{23}$-$B_{30})$-insulin, the liberation of which would have been indicative of a tryptic cleavage of the Arg-Gly bond, was not detectable even after an incubation period of 70 hr. The reaction mechanism of a similar "one-pot" synthesis involving the simultaneous incubation of porcine insulin and threonine methylester in the presence of trypsin and $H_2{}^{17}O$ was investigated using ^{17}O-NMR-spectroscopy by Markussen and Schaumburg.[14] The absence of a detectable ^{17}O signal in the newly formed human insulin methyl ester strongly supported the view that the enzymatic conversion took place via direct transpeptidation and without releasing an intermediate free DAI. Incorporation of ^{17}O into the carbonyl group of the Lys_{30}-residue would have been expected, if the transient DAI-trypsin complex had been hydrolytically deacylated prior to incorporation of the C-terminal threonine methylester.

In contrast to the above study,[14] Rose et al.[15] found, using a different solvent system and a different ester of threonine, that the trypsin-mediated conversion of porcine insulin into $(Thr(Bu^t)$-OBu^t-$B_{30})$-human insulin did not proceed by direct transpeptidation but rather via the hydrolized intermediate DAI. In the presence of water enriched with ^{18}O, the authors observed incorporation of ^{18}O into the Lys B_{29} carbonyl group of the human insulin ether ester. In consequence they were able to exclude a mechanism involving transpeptidation without prior hydrolysis.

In an alternative to the procedures discussed so far, Rose et al.[16] employed immobilized trypsin to convert porcine insulin in the presence of H-$Thr(Bu^t)$-OBu^t into $(Thr(Bu^t)$-OBu^t-$B_{30})$-human insulin. The authors were able to obtain the desired product within a comparatively short incubation period (2 hr) at a pH range between 5.3 and 5.6 and in an extraordinary high yield (99%).

Any possibility of hydrolyzing the Arg_{B22}-Gly_{B23} linkage during the coupling of DAI and H-Thr-OBu[t] can be eliminated by replacing trypsin with *Achromobacter lyticus* protease I as catalyst. The specificity of this protease is strictly confined to peptide bonds the carbonyl group of which is contributed by a lysine residue.[17] Morihara et al.[18] used this protease not only for the assembly of $(Thr$-OBu^t-$B_{30})$-insulin (yield: 83%) but also for the preceding step, i.e., the preparation of DAI from porcine insulin (Figure 2). The proteolytic reaction which gave DAI was carried out at pH 8.3, whereas the proteosynthetic, which provided the desired human insulin *t*-butylester took place at pH 6.5. Furthermore, Achromobacter protease I exhibited a coupling efficiency about ten times that of trypsin.[19] It has also been demonstrated that human insulin can be synthesized by a similar scheme but using immobilized Achromobacter protease I as catalyst for peptide bond formation.[20] The yield of the coupling step came to 74%. In contrast however catalysis by immobilized trypsin failed to yield the desired product. An additional study on Achromobacter protease I-mediated conversion of porcine insulin into human insulin was reported by Oka et al.[21] The authors presented a concept for the continuous preparation on an industrial scale of semisynthetic insulin. DAI and H-Thr-OBu[t] were coupled on a column prepacked with a polymeric carrier (SiO_2-polyglutamic

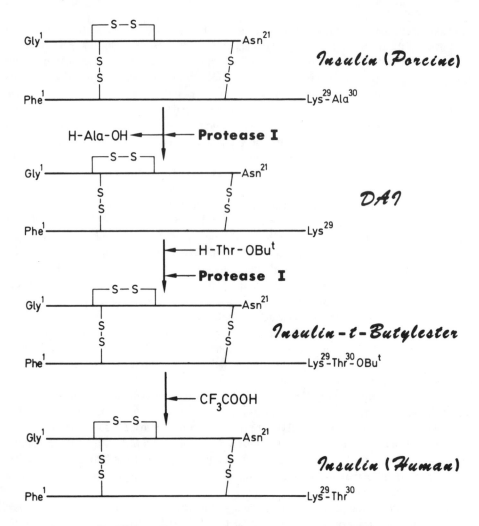

FIGURE 2.   Enzymatic conversion of porcine- into human insulin.

acid) to which protease I had been covalently linked. Thus the immobilized protease could be repeatedly used with increasing product yields.

Carboxypeptidase Y, an exopeptidase, was used by Breddam et al.[22] to transform porcine insulin into the human form by direct replacement of the original alanine B_{30} by a threonine residue. Considering the known transpeptidation potency of this exopeptidase,[23] this approach appeared to be a promising semisynthetic strategy. Indeed, the carboxypeptidase Y-catalyzed reaction between native, porcine insulin, and threonine amide furnished the desired human insulin as the first transpeptidation product (yield at pH 7.5: approximately 20%) this being subsequently deaminated by carboxypeptidase Y. However, the latter could not be purified to homogeneity, because des-Ala(B_{30})-insulin (DAI), which was formed concurrently by CPD-Y-mediated proteolysis, and the unreacted porcine insulin could not be resolved from the desired material. Furthermore, dimerization of the threonine derivative gave rise to another by-product, which was identified as human Thr_{30}-NH_2-insulin. In a supplementary study Breddam and Johansen observed that mercury halide derivatives of CPD-Y could catalyze the semisynthetic preparation of the human insulin B-chain more efficiently than native CPD-Y.[24]

Apart from the semisynthetic preparation of human insulin a variety of enzymatic semisyntheses have been performed which aim at well-defined modifications of the tripeptide segment B_{24-26} (Phe-Phe-Tyr) in the B-chain of insulin. This portion of the molecule is closely associated with the biological activity,[25] the aggregation,[26] and the receptor binding[27] of the hormone. In consequence, "mutations" of the primary structure in this important region may improve our understanding of how the amino acid sequence affects the functional and structural properties of insulin. Together with suitably modified, synthetic (B_{23-30})-octapeptides the desoctapeptide (B_{23-30})-insulin (DOI) (*vide supra,* this chapter) represents a useful starting point for the assembly of the desired insulin mutants via trypsin catalysis. Using a modification of the procedure developed by Inouye et al.,[3] two analogues of human insulin could be prepared in which either the phenylalanine residue B_{24} or N_{25} was replaced by a leucine residue.[28,29] Additionally either a tyrosine or an alanine residue could be introduced into position B_{25} in place of the phenylalanine residue,[30] and both the phenylalanine units — B_{24} and B_{25} — could be simultaneously substituted by leucine residues.[31] Starting from a DOI-derivative derived from bovine insulin, Inouye et al. obtained the (Ala B_{24}, Thr B_{30})-, (Ala B_{25}, Thr B_{30})-, and (Ala B_{26}, Thr B_{30}) analogs of bovine insulin via tryptic catalysis.[32] Furthermore, they reported the semisynthesis, via trypsin-catalysis, of (D-Phe B_{24})-, (D-Phe B_{25})-, and (D-Tyr B_{26})-human insulin starting from DOI and suitably "mutated" (B_{23-30}) octapeptides.[33]

In addition to the semisynthetic insulin analogues described above, the following truncated insulin derivates were prepared in the presence of trypsin from DOI and various synthetic peptides: a destetrapeptide $(B_{27}\text{-}B_{30})$ insulin, a destetrapeptide $(B_{27}\text{-}B_{30})$, (Phe B_{26}) insulin, the truncated B-chain of both of these mutants being C-terminally methylated,[30] a despentapeptide (B_{26-30})-insulin-B_{25}-amide[34] and a deshexapeptide (B_{25-30})-insulin.[35]

The functional and structural properties of several insulin "mutants", carrying modifications in the B_{24-26} region[28,29,31,32] as determined by biological and optical measurements[32] indicate that the residues Phe B_{24} and to a lesser extent B_{25}, make an important contribution to the mechanism of insulin action. "Mutations" directed toward this portion of the B-chain significantly reduce the receptor binding potency of insulin, whereas the replacement of Tyr B_{26} by and alanine residue hardly influences its biological effectiveness. Beyond that, the residue Phe B_{24} appears to be particularly important for antibody recognition, since both the Leu$_{24}$- and the Ala B_{24}-analog display a considerable decline in the affinity for antibodies raised against unmodified insulin. As can be deduced from CD measurements, the replacement of Phe B_{24} by Leu B_{24} lowers the capacity of insulin to form aggregates. The reduced potency of the above insulin analogues to bind antibodies can therefore be attributed to the finding that both the Ala B_{24}- and the Leu$_{24}$ "mutants" lack the β-structure which is a predominant feature of the normal insulin dimer.

The tryptic semisyntheses of insulin discussed so far employ intermolecular reactions in which peptide bond formation is encouraged by the presence of a considerable molar excess of the respective amine component. A quite different approach has recently been described by Chu et al.[36] These authors succeeded in restoring the trypsin-split internal bond between the residues Arg B_{22} and Gly B_{23} of an $N^{\alpha A1}$, $N^{\epsilon B29}$-(carbonyl-*bis*-methionyl) insulin in an unimolecular reaction. Thus after tryptic cleavage of the Arg-Gly bond the molecule was prevented from falling apart by the presence of the carbonyl-*bis*-methionyl cross-link between the A- and B-chains. As a result, the subsequent synthetic step represented an intramolecular process. Although the maintenance of the original position of the (B_{23-30}) octapeptide in the native molecule cannot be guaranteed, the prospects of a conformationally driven covalent synthesis were favored due to the steric proximity of the intramolecular reactants. The yield from tryptic restoration of the Arg-Gly bond (60 to 67%) was remarkable in view of the fact that the concentration of the reactants was significantly reduced (ca. 50-fold in comparison to the foregoing tryptic syntheses) and the molar ratio of the reactants was necessarily 1:1. It was an intramolecular reaction, too, which gave rise to the trypsin-controlled formation

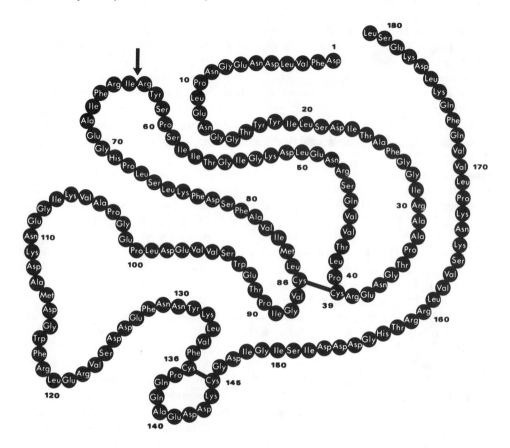

FIGURE 3. Primary structure of soybean trypsin inhibitor.[38] (The arrow designates the trypsin-sensitive peptide bond of the reactive site).

of a single chain DAI.[37] The molecular mechanism of this process is assumed to be as follows: First, upon incubation of porcine insulin with trypsin, the Ala B_{30} residue is removed and a covalent acyl-protease complex is formed via an ester bond between the carboxyl group of Lys B_{29} and the hydroxylic function of the active-site serine residue of trypsin. The next step involves the aminolytic cleavage of the acyl-enzyme-complex by the α-amino group of the Gly A_1 residue resulting in the formation of a novel Lys B_{29}-Gly A_1 peptide bond. The product yield (13%) is small relative to the foregoing intramolecular reaction. Possibly, the sterical constraints imposed by the covalent insulin-trypsin complex favored the contact between Arg B_{22} and Gly B_{23} rather than that between Lys B_{29} and Gly A_1.

III. PROTEASE INHIBITORS

Pioneering studies on protease-catalyzed protein semisynthesis of "enzymatically mutated" soybean trypsin inhibitors (Kunitz) (STI) have been performed by Laskowski, Jr. and his associates. STI belongs to the subclass of protein protease-inhibitors which act as strong, competitive inhibitors of serine proteases. Their inhibitory potential is based on their capability to form strong complexes with their respective protease. Thus for example the affinity of trypsin for its inhibitor greatly exceeds that for its common substrates. The STI-molecule consists of a single-chain polypeptide which contains two disulfide bridges and 181 amino acid units, among them 9 arginine and 10 lysine residues (Figure 3).[38] STIs reactive-site peptide bond between the residues Arg_{63} and Ile_{64} can be selectively cleaved

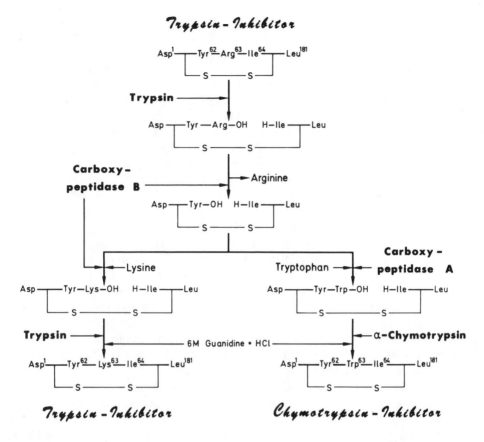

FIGURE 4. Protease-catalyzed exchange of an amino acid residue located at the active site of trypsin inhibitor (STI).

in the presence of trypsin.[39] However, despite the split Arg_{63}-Ile_{64} bond, the resulting "modified" inhibitor (STI*) displays the same inhibitory activities as the intact molecule ("virgin" inhibitor). Obviously, the tertiary structure in the direct environment of the reactive site, which is stabilized both by the disulfide bond (Cys_{39}-Cys_{86}) and by a variety of secondary, noncovalent forces,[40] is not significantly changed by the tryptic cleavage.

Using an elegant semisynthetic concept, that involved the exploitation of both the proteolytic and proteosynthetic potentials of trypsin and carboxypeptidase B (CPD-B), the internal Arg_{63} residue of STI could be replaced by a lysine residue.[41] As the newly formed semisynthetic (Lys_{63})-STI is also a strong inhibitor of trypsin, this "site-directed" mutagenesis proved the formal equivalence of the arginine and the lysine residue in the active site of STI.

The amino acid exchange (Arg_{63} by Lys) was accomplished by the following series of protease-catalyzed steps (Figure 4): First, the "virgin inhibitor" (STI) was incubated in the presence of trypsin (pH 3.75) to produce the modified inhibitor (STI*) via selective cleavage of the Arg_{63}-Ile_{64} bond. Subsequent treatment with carboxypeptidase B (pH 7.6) led to the removal of the Arg_{63} residue. The resulting des-(Arg_{63}) inhibitor bound hardly at all to trypsin and had consequently lost its inhibitory properties. The incubation of the inactive des-(Arg_{63})-STI and free lysine with carboxypeptidase B in the presence of trypsin (pH 6.7) enabled the covalent introduction of a lysine residue into the original site of Arg_{64}. This novel, "modified" (Lys_{63}) inhibitor still lacked a covalent Lys_{63}-Ile_{64} bond, but it nevertheless formed a strong complex with trypsin. The dissociation of this complex in the presence of

6 M guanidine hydrochloride finally furnished the "virgin"-(Lys$_{63}$)-trypsin inhibitor with an intact Lys$_{63}$-Ile$_{64}$ peptide bond. The semisynthetic strategy of Sealock and Laskowski[41] illustrates some factors that ensured an efficient Arg$_{63}$-Lys exchange.

Although the CPD-B-catalyzed introduction of a lysine residue was driven by mass action in the presence of a large molar surplus of free lysine (700-fold), the reaction was far from going to completion. Hence, an additional driving force was required to obtain a reasonable product yield. Consequently, the thermodynamically unfavorable process of peptide bond formation $K_{syn} = 10^{-2}M^{-1}$ was coupled to the highly exergonic formation of the complex between trypsin and its inhibitor ($K_{ass} = 10^{11}M^{-1}$).[42] As a result, the newly formed "modified" (Lys$_{63}$) inhibitors (STI*) were immediately "sequestrated" by trypsin acting as a "molecular trap", and the overall equilibrium was shifted overwhelmingly toward proteosynthesis ($K*_{syn} = 10^{9}M^{-2}$). To obtain the free "virgin" (Lys$_{63}$)-trypsin inhibitor the Lys$_{63}$-Ile$_{64}$ bond must be formed and the resulting complex cleaved. However, if this was allowed to occur under thermodynamic control, the partition between free (Lys$_{63}$)-STI and free (Lys$_{63}$)-STI* would lead preferentially to the free "modified" inhibitor being released, because the equilibrium lies mainly on the side of (Lys$_{63}$)-STI*. At pH 2, for example, the (Lys$_{63}$)-STI would merely represent of 1% of the overall yield. On the other hand, if a sudden cleavage of the complex can be accomplished by rapidly lowering the pH[43] or — as done in the present case — by the presence of 6 M guanidine hydrochloride, the initial distribution between free (Lys$_{63}$)-STI and free (Lys$_{63}$)-STI* is kinetically controlled, i.e., it depends solely on the ratio of the dissociation rate constants of both inhibitor forms. Since the dissociation rate of STI (intact peptide bond) largely exceeds that of STI* (split peptide bond), the virgin (Lys$_{63}$)-inhibitor is predominantly formed.

By using a strategy similar to that described above, Leary and Laskowski, Jr.[42,44] succeeded in replacing Arg$_{63}$ by a tryptophan residue (cf. Figure 4). Starting from the previously mentioned des-(Arg$_{63}$)-trypsin inhibitor Trp$_{63}$ was introduced via CPD-A catalysis in the presence of chymotrypsin. The synthesis of the peptide bond was again promoted by the formation of a complex which, in this case, was composed of the newly synthesized "modified" (Trp$_{63}$)-inhibitor and chymotrypsin. Kinetically controlled dissociation in 6 M guanidine hydrochloride provided the virgin (Trp$_{63}$)-inhibitor containing an intact Trp$_{63}$-Ile$_{64}$ peptide bond. This molecule strongly inhibited chymotryptic activities, while it showed no inhibitory effect on tryptic activity. The results of the above "site-directed mutageneses" indicate that the inhibitory specificity is governed by primarily the chemical nature of the amino acid residue at position 63 of the inhibitor. However, the replacement of Arg$_{64}$ by a phenylalanine residue resulted in an inhibitor, albeit very weak, of both tryptic and chymotryptic activities,[42] although chymotrypsin inhibitors having a phenylalanine residue at their active site do exist in nature.

The protease-catalyzed substitution of the specificity-controlling Lys$_{15}$ residue of bovine trypsin-kallikrein inhibitor (Kunitz) (TKI) by respectively arginine, phenylalanine, and tryptophan, has been described by Jering and Tschesche.[45,46] The TKI-molecule consists of a single-chain polypeptide composed of 58 amino acid residues. Its primary structure and the position of the three intrachain disulfide bridges was elucidated by Kassel and Laskowski, Sr.[47,48] One of these disulfide bridges — Cys$_{14}$-Cys$_{38}$ — is located close to the active-site which is constituted by the Lys$_{15}$-Ala$_{16}$ peptide bond. By analogy to STI*, a modified but still active inhibitor (TKI*) can be prepared in which the peptide bond Lys$_{15}$-Ala$_{16}$ has been hydrolyzed. (The rather complicated preparation of TKI* was described by Jering and Tschesche.[49,50]) Starting from TKI*, the replacement of Lys$_{15}$ by Arg$_{15}$, Trp$_{15}$, and Phe$_{15}$ was accomplished via a series of protease-catalyzed steps[45,46] similar to those described for the modification of STI.[41,42,44] The inhibitory capacity vs. trypsin, plasmin, and kallikrein of the semisynthetic (Arg$_{15}$)-TKI equalled that of native TKI. Meanwhile the (Phe$_{15}$)- and the (Trp$_{15}$) homologs displayed weak inhibitory activity toward trypsin and kallikrein, but were powerful inhibitors of chymotrypsin. The inhibitory specificity of (Trp_{15})-TKI is comparable to that of (Trp_{63})-STI, whereas the strong inhibitory potential of (Phe_{15})-TKI did not correspond to that of (Phe_{63})-STI, which was a rather weak inhibitor of chymotrypsin.

The enzymatic restoration of an *a priori* hydrolyzed peptide bond at the active site of a protease-inhibitor was reported by Ardelt and Laskowski, Jr.[51] In this study the inhibitor was represented by the third domain of the turkey ovomucoid which was obtained by limited hydrolysis of the native ovomucoid.[52] (The inhibitory potency of the native molecule is inferior to that of its subunit.) The serine protease aspergillopeptidase B from *Aspergillus oryzae* was used as catalyst for the resynthesis of the Leu_{18}-Glu_{19}-peptide bond in the active site of the inhibitor. In contrast to the inhibitors described so far, the ovomucoid domain inhibits the activity of its "prey", in the present case aspergillopeptidase B, only to a minor extent. The equilibrium constant for the formation of the inhibitor-protease complex ($1.2 \times 10^7 M^{-1}$) is several orders of magnitude smaller than those observed for trypsin inhibitors. In addition, thermodynamic studies revealed that the equilibrium constants for the hydrolysis and the synthesis of the above-mentioned peptide bond are roughly equal.

The synthetic studies on ovomucoid protease inhibitors have led to the enzymatic preparation of an inhibitor chimera, which can be assembled from ovomucoid fragments of different avian origin.[53] The strategy behind this approach is summarized as follows (see also Figure 5). First, the third domains of both turkey- and Gambel's quail ovomucoid, the primary structures of which closely resemble each other, were modified by cleaving the Leu_{18}-Glu_{19} bonds with protease B from *Streptomyces griseus*. Subsequently the disulfide bridges were split by means of an SH reagent (dithiothreitol) to separate the respective peptide fragments. These were then treated with oxidized glutathione to give the respective mixed disulfides. The oxidized peptide fragments corresponding to residues 1 to 18 of turkey ovomucoid and residues 19 to 56 of the quail ovomucoid were then combined in the presence of cysteine in order to reform the disulfide bonds. The outcome of this procedure was a hybridized inhibitor whose subunits were indeed connected via two disulfide bridges. However to complete the synthesis, the active-site Leu_{18}-Glu_{19} peptide bond had to be formed. This was achieved in the presence of either proteinase K from *Tritiachium album* or of aspergillopeptidase B and finally provided the intact inhibitor chimera; though the aspergillopeptidase B-catalyzed reaction gave the truncated des-Leu_1-inhibitor. The binding behavior v5. chymotrypsin of the semisynthetic inhibitor hybrid paralleled that of the native inhibitors.

IV. NUCLEASES

Bovine pancreatic ribonuclease S (RNase-S), which over the years has become a "classic" among the semisynthetic systems, is the product of a subtilisin-controlled, selective cleavage of the Ala_{20}-Ser_{21} bond in the native ribonuclease A.[54] The two resulting subunits of RNase-S — the so-called S-peptide (RNase-S_{1-20}) and the S-protein (RNase-S_{21-124}) — form a noncovalent complex which maintains the original enzymatic activity. Homandberg and Laskowski, Jr.[55] demonstrated the reversibility of the abovementioned, proteolytic process; i.e., RNase-A could be recovered from RNase-S by enzymatically restitching the split Ala_{20}-Ser_{21} bond (cf. Figure 6). The product yield which was only 4.3% in an aqueous solution could be improved by the addition of increasing amounts of glycerol. In the presence of 90% (v/v) glycerol the restored RNase-A could be obtained in 50% yield. The equilibrium constant for proteosynthesis (K_{syn}), estimated by the authors to be $5 \times 10^{-2} M^{-1}$ in an aqueous environment having a pH range of 6.2 to 6.5, was roughly unity when the reaction medium contained 90% glycerol. Consequently, the equilibrium constant for the hydrolysis (K_{hyd}) of the Ala_{20}-Ser_{21} bond in an aqueous solution amounts to 20; a surprisingly low value which however, according to the authors, can be explained by the restricted motional freedom

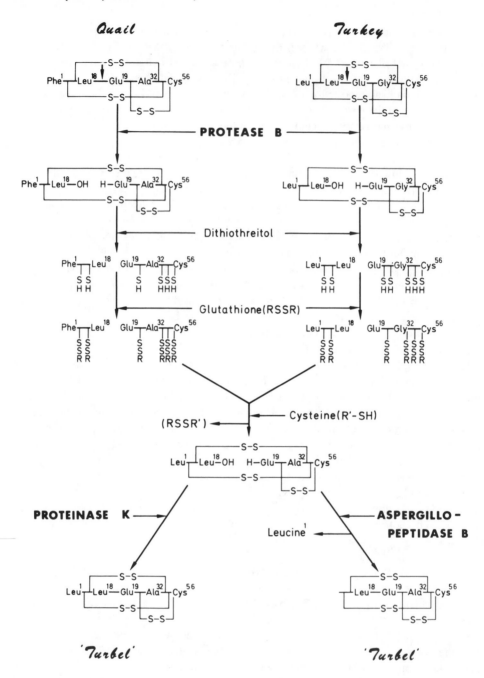

FIGURE 5.   Enzymatic construction of a protease-inhibitor chimera, assembled from ovomucoid fragments of a quail and a turkey.

of the amino acid residues in the neighborhood of the above peptide bond. In fact the product yield of 4.3% does significantly exceed those values which should be expected for the protease-mediated synthesis of peptide bonds in absence of added co-solvents. Thus, the subtilisin BNP′-catalyzed restoration of the Ala_{20}-Ser_{21} peptide bond was, at least in part, conformationally controlled, i.e., it was favored by the steric proximity of the residues Ala_{20} and Ser_{21} within the S-peptide-S-protein complex. The major contribution to the resynthesis, however, was provided by the addition of glycerol. By reducing both the acidity of the

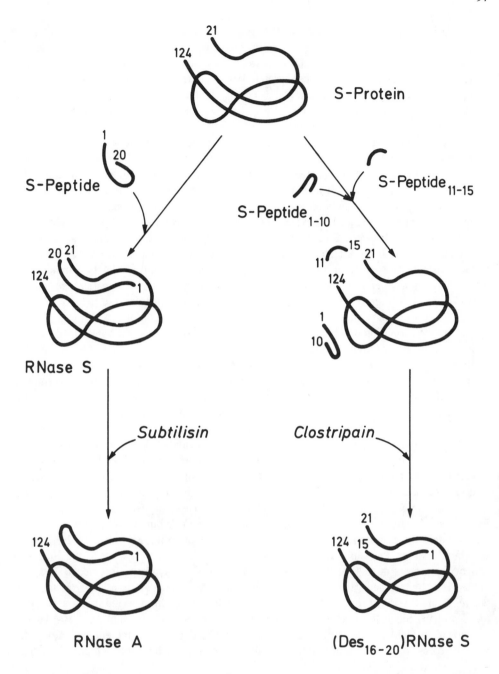

FIGURE 6. Schematic representation of the recovery of RNase-A, and the semisynthesis of a "deletion mutant" of RNase-S.

carboxyl group of Ala_{20} and the water concentration *(vide supra)* the organic co-solvent was able to supply the main driving force for the recovery of RNase-A.

Strictly speaking, this enzymatic approach to the restoration of the native RNase-A is not truly semisynthetic, because the separated subunits were both of natural origin. An alternative enzyme-mediated approach to the preparation of RNase-A utilizes a genuinely semisynthetic procedure. Starting from a knowledge of the experimental conditions prevailing during the study described above,[55] Homandberg et al.[56] combined crude synthetic S-peptide with native

S-protein in the presence of subtilisin BPN' (cf. Figure 6) in a solvent system containing 10% water and 90% glycerol. In this system, the prospect of restoring an active RNase-A containing the authentic S-peptide selected from the crude mixture of synthetic peptides is excellent for two reasons. First, the S-protein demonstrates a clear preference for complex formation with the authentic S-peptide and second, the structural specificity of the protease means that a peptide bond can only be formed between the S-protein and an S-peptide of the correct 1 to 20 amino acid sequence. A 30% yield of active RNase-A could be obtained (in the presence of native S-peptide a similar semisynthetic procedure yielded 50% RNase-A; see above) even though the purity of the crude solid-phase-derived synthetic peptide only comes to about 30 to 50%. The facility of the method of "bifunctional purification" in selecting the correct peptide from a crude mixture is emphasized by the fact that the by-products of chemical syntheses are often very closely related to the main product.

The preceding studies on the covalent reconstitution of RNase-A[55,56] were based on the complex formation between S-peptide and S-protein. In such a complex the arrangement of the reactive groups is such as to ensure both their spatial proximity and their accessibility to the protease. However, in other reports on the preparation of a "deletion mutant" of RNase-S, Komoriya et al.[57] and Homandberg et al.[58] described the enzymatic condensation of two *ex arte* subfragments of the S-peptide which were ordinarily unable to associate with each other or to form, either individually or in combination, a complex with the S-protein. Following the formation of a peptide bond between these subfragments, however, the newly formed condensation product could then noncovalently bind to the S-protein (cf. Figure 6). Unlike the protease-catalyzed semisynthesis of the modified trypsin inhibitor, which was mediated by an autonomous "extrinsic" factor, namely by trypsin,[41] the present semisynthesis is driven by an "intrinsic" trap, namely by the provision of an integral part of the prospective product. The nonassociating fragments, which corresponded to the sequences (1 to 10) and (11 to 15) of the S-peptide, were enzymatically coupled in the presence of natural S-protein. The peptide bond forming step was catalyzed by clostripain (clostridiopeptidase B) a protease from *Clostridium histolyticum,* the primary specificity of which is directed toward an arginine residue at the P_1-substrate site.[59] This enzyme was chosen as catalyst because of its selective action on the arginine residue occupying the C-terminal position of the S-peptide subfragment (1 to 10). The use of trypsin, for example, would jeopardize not only the Lys_7-Phe_8 linkage within the S-peptide but also some further peptide bonds within the S-protein. The yield of the novel "deletion mutant" des-(16-20)-RNase-S, which displays the same activity as RNase-S, was 80% with respect to the S-protein which was present at 1/20 of the molar concentration of the synthetic fragments. Thus 4% of the synthetic fragments, which were present in equimolar concentrations, were condensed, whereas only 0.05% of the condensed product would have been expected in the absence of the S-protein. Since the reaction was carried out in an aqueous medium, the 80-fold improvement in yield can be attributed exclusively to the coupling of the energetically unfavorable process of peptide bond formation with the energetically favorable complex formation, rather than to any solvent effect.

In contrast to the above, a study on the trypsin-catalyzed condensation of nuclease-T fragments[60] illustrates one of the problems that can sometimes result from the coupling of the thermodynamically unfavorable synthesis of a peptide bond with a highly exergonic process. Limited tryptic proteolysis of the staphylococcal nuclease, which is a single-chain polypeptide of 149 amino acid residues, yields as the first cleavage product a fully active nuclease-T which however lacks the N-terminal pentapeptide. Additional tryptic treatment results in peptide fragments corresponding to the sequences (6 to 48), (49 to 149), and 50 to 149) of the native nuclease. These tryptic cleavage products generate a roughly equal mixture of the noncovalent complexes, nuclease-T_a (6 to 48)-(49 to 149) and nuclease-T_b (6 to 48)-(50 to 149), each of which possesses a residual activity corresponding to approximately 8% of that of the original nuclease.[61,62] Incubation of either the nuclease-T_a- or the T_b-complex at pH 6.2 in the presence of trypsin in a solvent system containing 90% (v/v) glycerol furnished the covalent forms (yield: $\sim$ 30%, in both cases) which were recognized as des-(Lys_{49})-nuclease-T. This deleted single chain protein showed less than 0.5% of the original nuclease-T activity.

The behavior of the nuclease-T_b (6 to 48)-(50 to 149) complex in the presence of trypsin was predictable because the K_{syn} of the peptide bond forming step in 90% glycerol is about 0.4. However, the formation via trypsin-catalysis of des-(Lys_{49})-nuclease-T from nuclease-T_a (6 to 48)-(49 to 149) was unexpected. In fact the reaction pathway leading to this "deletion mutant" probably proceeds as follows: first, the desired conversion of the nuclease-T_a complex into the nuclease-T (6 to 149) occurs; i.e., the Lys_{48}-Lys_{49} peptide bond is restored. However, both the newly formed peptide bond and the Lys_{49}-Gly_{50} bond are then cleaved via tryptic proteolysis. As a consequence, the Lys_{49} residue is excised to yield the above-mentioned nuclease-T_b complex which is subsequently converted via tryptic proteosynthesis into des-(Lys_{49})-nuclease-T. Thus the resynthesis of nuclease-T (6 to 149) finally fails in this case by virtue of its coupling to a highly exergonic but undesired process, the almost irreversible excision of the Lys_{49} residue. However, this problem can be circumvented by using a modified nuclease (6 to 49) fragment, in which Lys_{48} was replaced by a glycine residue.[63] In the present case the "natural" nuclease-T-(50 to 149) fragment and a tenfold molar excess of the synthetic (Gly_{48})-nuclease-T-(6 to 49) fragment were incubated with trypsin in 90% (v/v) glycerol to yield 20% of (Gly_{48})-nuclease-T with the Lys_{49}-Gly_{50} bond intact. The activity of this "mutated" nuclease-T was roughly equal to that of the native nuclease.

V. MISCELLANEOUS

Promising results have also been reported for the protease-catalyzed semisynthesis of cytochrome c analogs.[64] Being an integral part of the respiratory chain, cytochrome c plays a crucial role in electron transport. A covalently bound heme moiety confers the characteristic red color on the electron transferring protein. Cytochrome c (from horse heart) is composed of 104 amino acid residues, including 19 lysine and 2 arginine residues whose basic charges are not fully counterbalanced by 15 aminodicarboxylic acids,[65] the protein therefore being basic in character. The selective, tryptic cleavage of the Arg_{38}-Lys_{39} peptide bond of the N^ϵ-protected cytochrome c generates the so-called ferrous "heme" fragment (1 to 38) and the apo-fragment (39 to 104).[66] These fragments can form a noncovalent complex which displays reduced biological activity. The trypsin-catalyzed synthesis of an α-peptide bond between the natural heme fragment, the side-chain amino functions of which were protected by Msc groups, and the synthetic model dipeptide H-Orn(δ-Msc)-Trp-NHCH$_3$ was reported by Westerhuis et al.[67] The peptide bond forming step, which was performed according to a procedure described by Morihara et al.[10] yielded 37% of the condensed product. In this system when the lysine side chains were not protected, the heme fragment instead underwent partial tryptic hydrolysis. In contrast to this, native cytochrome c could be restored in up to 30% yield from the unprotected heme- and apo-fragments in the presence of the arginine-specific protease clostripain.[56,68] In this case the addition of 90% (v/v) glycerol greatly reduced the thermodynamic barrier to peptide bond formation. In a different study Proudfoot and Wallace[69] accomplished the back-conversion of the complexing fragments (1 to 38) and (39 to 104) to native cytochrome c (ca. 20% yield) in the absence of enzyme simply by keeping the reactants in 90% glycerol. In fact, the above yield could not be significantly improved in the presence of clostripain. In the absence of glycerol and clostripain no condensed product was formed at all.

The positive effect of organic co-solvents (90% glycerol) on the peptide bond forming step was also exploited when Graf and Li[70] demonstrated the reconstitution of the Arg_{134}-Thr_{135} peptide linkage of human somatotropin (hGH), a 191-residue protein. The protease thrombin, which had previously been used to selectively cleave the Arg_{134}-Thr_{135} bond was again used to condense the two 134- and 57-residue fragments at pH 6.0 (yield: ca. 20%). Presumably the peptide bond formation was favored not only by the presence of glycerol but also by the steric proximity and the proper conformation of the peptide fragments. This may be particularly the case for fragments as large as the above hGH subunits which represent the largest protein fragments enzymatically resynthesized so far.

The trypsin-catalyzed preparation of a modified human hemoglobin (Hb)A has been reported by Nagai et al.[71] As compared to native (Hb)A, the modified (Hb)A exhibited an increased affinity for oxygen and a reduced alkaline Bohr effect. The incubation of native (Hb)A and $H-Gly-NH_2$ with trypsin generated a hemoglobin, $Gly-NH_2(142\alpha)$-Hb, in which the glycine amide was covalently attached via a peptide bond to the C-terminal Arg_{141} residue of the α-subunit. Since the incorporation of glycine amide was performed in the absence of organic co-solvents, the peptide bond forming step (yield: ca. 50%) was driven predominantly by mass action, the $H-Gly-NH_2$ being present in approximately 3000-fold molar excess. The altered oxygen binding characteristics of $Gly-NH_2(142\alpha)$-Hb are explained by the fact that in native desoxy(Hb)A the guanidino group of $Arg_{141\alpha}$ forms an anion-mediated salt bridge with the α-amino group of $Val_{1\alpha}$ of the opposite α-chain thereby stabilizing the reduced state.[72] The addition of a glycine amide to the C-terminal Arg_{141} residue prevents this interaction thereby enabling oxygen to bind more easily albeit with a loss of the cooperativity of oxygen binding normally mediated by the breakage of the interchain salt bridge.

REFERENCES

1. **Gross, E. and Meienhofer, J.,** The peptide bond, in *The Peptides: Analysis, Synthesis, Biology,* Vol. I, Gross, E. and Meienhofer, J., Eds., Academic Press, New York, 1979, 1.
2. **Nicol, D. S. H. W. and Smith, L. F.,** Amino-acid sequence of human insulin, *Nature (London),* 187, 483, 1960.
3. **Inouye, K., Watanabe, K., Morihara, K., Tochino, K., Kanaya, T., Emura, J., and Sakakibara, S.,** Enzyme-assisted semisynthesis of human insulin, *J. Am. Chem. Soc.,* 101, 751, 1979.
4. **Ruttenberg, M. A.,** Human insulin: facile synthesis by modification of porcine insulin, *Science,* 177, 623, 1972.
5. **Obermeier, R. and Geiger, R.,** A new semisynthesis of human insulin, *Hoppe Seyler's Z. Physiol. Chem.,* 357, 759, 1976.
6. **Sanger, F. and Tuppy, H.,** The amino-acid sequence in the phenylalanyl chain of insulin, *Biochem. J.,* 49, 481, 1951.
7. **Young, J. D. and Carpenter, F. H.,** Isolation and characterization of products formed by the action of trypsin on insulin, *J. Biol. Chem.,* 236, 743, 1961.
8. **Harris, J. J.,** The use of carboxypeptidase for the identification of terminal carboxyl groups in polypeptides and proteins. Asparagine as a C-terminal residue in insulin, *J. Am. Chem. Soc.,* 74, 2944, 1952.
9. **Schmitt, E. W. and Gattner, H.-G.,** Verbesserte Darstellung von Des-alanyl B_{30}-insulin, *Hoppe Seyler's Z. Physiol. Chem.,* 359, 799, 1978.
10. **Morihara, K., Oka, T., and Tsuzuki, H.,** Semisynthesis of human insulin by trypsin-catalyzed replacement of Ala-B 30 by Thr in porcine insulin, *Nature (London),* 280, 412, 1979.
11. **Gattner, H.-G., Danho, W., and Naithani, V. K.,** Enzyme catalyzed semisynthesis with insulin derivatives, in *Proc. 2nd Int. Insulin Symp.,* Aachen, W. Germany, 1979, Brandenburg, D. and Wollmer, A., Eds., Walter de Gruyter, Berlin, 1980, 117.
12. **Morihara, K., Oka, T., Tsuzuki, H., Inouye, K., Tochino, Y., Kanaya, H., Masaki, T., Soejima, M., and Sakakibara, S.,** *Peptide Chemistry 1979,* Yonehara, H., Ed., Protein Research Foundation, Osaka, Japan, 1980, 113.
13. **Jonczyk, K. A. and Gattner, H.-G.,** Eine neue Semisynthese des Humaninsulins. Tryptisch-katalysierte Transpeptidierung von Schweineinsulin mit L-Threonin-*tert*-butyl-ester, *Hoppe Seyler's Z. Physiol. Chem.,* 362, 1591, 1981.

14. **Markussen, J. and Schaumburg, K.,** Reaction mechanism in trypsin catalyzed synthesis of human insulin studied by ¹⁷O-NMR spectroscopy, in *Peptides 1982, 17th Eur. Peptide Symp.,* Bláha, K. and Maloň, P., Eds., Walter de Gruyter, Berlin, 1983, 387.

15. **Rose, K., Gladstone, J., and Offord, R. E.,** A mass-spectrometric investigation of the mechanism of the semisynthetic transformation of pig insulin into an ester of insulin of human sequence, *Biochem. J.,* 220, 189, 1984.

16. **Rose, K., de Pury, H., and Offord, R. E.,** Rapid preparation of human insulin and insulin analogues in high yield by enzyme-assisted semi-synthesis, *Biochem. J.,* 211, 671, 1983.

17. **Masaki, T., Nakamura, K., Isono, M., and Soejima, M.,** A new proteolytic enzyme from *Achromobacter lyticus* M 497-1, *Agric. Biol. Chem.,* 42, 1443, 1978.

18. **Morihara, K., Oka, T., Tsuzuki, H., Tochino, Y., and Kanaya, T.,** Achromobacter protease I-catalyzed conversion of porcine insulin into human insulin, *Biochem. Biophys. Res. Commun.,* 92, 396, 1980.

19. **Morihara, K. and Oka, T.,** Enzymatic semisynthesis of human insulin by transpeptidation method with Achromobacter protease: comparison with the coupling method, in *Peptide Chemistry 1982,* Sakakibara, S., Ed., Protein Research Foundation, Osaka, Japan, 1983, 231.

20. **Muneyuki, R., Oka, T., and Morihara, K.,** Peptide synthesis using enzyme as synthetic catalyst — synthesis of new water-soluble ester substrates and enzyme immobilization-, *Nippon Kagaku Kaishi,* 1336, 1983.

21. **Oka, T., Muneyuki, R., and Morihara, K.,** Enzymatic semisynthesis of human insulin: a proposed procedure using immobilized enzyme, in *Peptides, Structure and Function, Proc. 8th Am. Peptide Symp.,* Hruby, V. J. and Rich, D. H., Eds., Pierce Chem., Rockfort, Ill., 1983, 199.

22. **Breddam, K., Widmer, F., and Johansen, J. T.,** Carboxypeptidase Y catalyzed C-terminal modification in the B-chain of porcine insulin, *Carlsberg Res. Commun.,* 46, 361, 1981.

23. **Breddam, K., Widmer, F., and Johansen, J. T.,** Carboxypeptidase Y catalyzed transpeptidations and enzymatic peptide synthesis, *Carlsberg Res. Commun.,* 45, 237, 1980.

24. **Breddam, K., and Johansen, J. T.,** Semisynthesis of human insulin utilizing chemically modified carboxypeptidase Y, *Carlsberg Res. Commun.,* 49, 463, 1984.

25. **Weitzel, G., Bauer, F., and Eisele, K.,** Structure and activity of insulin. XIV. Further studies on the three-step-increase in activity due to the aromatic amino acids B24-26 (-Phe-Phe-Tyr-), *Hoppe Seyler's Z. Physiol. Chem.,* 357, 187, 1976.

26. **Blundell, T. L., Dodson, G. G., Hodgkin, D. C., and Mercola, D. A.,** Insulin: the structure in the crystal and its reflection in chemistry and biology, *Adv. Prot. Chem.,* 26, 279, 1972.

27. **Pullen, R. A., Lindsay, D. G., Wood, S. P., Tickle, I. J., Blundell, T. L., Wollmer, A., Krail, G., Brandenburg, D., Zahn, H., Gliemann, J., and Gammeltoft, S.,** Receptor-binding region of insulin, *Nature (London),* 259, 369, 1976.

28. **Gattner, H.-G., Danho, W., Behn, C., and Zahn, H.,** The preparation of two mutant forms of human insulin, containing leucine in position B 24 or B 25, by enzyme-assisted synthesis, *Hoppe Seyler's Z. Physiol. Chem.,* 361, 1135, 1980.

29. **Tager, H., Thomas, M., Assoian, R., Rubenstein, A., Saekow, M., Olefski, J., and Kaiser, E. T.,** Semisynthesis and biological activity of porcine (Leu B₂₄) insulin and (Leu B₂₅) insulin, *Proc. Natl. Acad. Sci. U.S.A.,* 77, 3181, 1980.

30. **Gattner, H.-D., Danho, W., Knorr, R., Naithani, V. K., and Zahn, H.,** Trypsin catalyzed peptide synthesis: modification of the B-chain C-terminal region of insulin, in *Peptides 1980, Proc. 16th Eur. Peptide Symp.,* Brunfeldt, K., Ed., Scriptor, Kopenhagen, 1981, 372.

31. **Jonczyk, K. A., Keefer, L. M., Naithani, V. K., Gattner, H.-D., De Meyts, P., and Zahn, H.,** Preparation and biological properties of (Leu B₂₄, Leu B₂₅) human insulin, *Hoppe Seyler's Z. Physiol. Chem.,* 362, 557, 1981.

32. **Inouye, K., Watanabe, K., Tochino, Y., Kobayashi, M., and Shigeta, Y.,** Semisynthesis and properties of some insulin analogs, *Biopolymers,* 20, 1845, 1981.

33. **Inouye, K., Watanabe, K., Tochino, Y., Kobayashi, M., Shigeta, Y.,** Semisynthesis of human insulin analogues containing a D-amino acid residue in position B 24, B 25, or B 26, in *Peptide Chemistry 1982,* Sakakibara, S., Ed., Protein Research Foundation, Osaka, Japan, 1983, 277.

34. **Fischer, W. H., Saunders, D., Brandenburg, D., Wollmer, A., and Zahn, H.,** A shortened insulin with full in vitro potency, *Hoppe Seyler's Z. Physiol. Chem.,* 366, 521, 1985.

35. **Cao, Q. P., Cui, D. F., and Zhang, Y. S.,** Enzymatic synthesis of deshexapeptide insulin, *Nature (London),* 292, 774, 1981.

36. **Chu, S.-C., Wang, C.-C., and Brandenburg, D.,** Intramolecular enzymatic peptide synthesis: trypsin-mediated coupling of the peptide bond between B22-arginine and B23-glycine in a split crosslinked insulin, *Hoppe Seyler's Z. Physiol. Chem.,* 362, 647, 1981.

37. **Markussen, J., Jørgensen, K. H., Sørensen, A. R., and Thim, L.,** Single chain des-(B 30) insulin, *Int. J. Peptide Protein Res.,* 26, 70, 1985.

38. **Koide, T. and Ikenaka, T.,** Studies on soybean trypsin inhibitors. III. Amino-acid sequence of the carboxyl-terminal region and the complete amino-acid sequence of soybean trypsin inhibitor (Kunitz), *Eur. J. Biochem.,* 32, 417, 1973.

39. **Ozawa, K. and Laskowski, M., Jr.,** The reactive site of trypsin inhibitor, *J. Biol. Chem.,* 241, 3955, 1966.

40. **Sweet, R. M., Wright, H. T., Janin, J., Clothia, C. H., and Blow, D. M.,** Crystal structure of the complex of porcine trypsin with soybean trypsin inhibitor (Kunitz) at 2.6-Å resolution, *Biochemistry,* 13, 4212, 1974.

41. **Sealock, R. W. and Laskowski, M., Jr.,** Enzymatic replacement of the arginyl by a lysyl residue in the reactive site of soybean trypsin inhibitor, *Biochemistry,* 8, 3703, 1969.

42. **Laskowski, M., Jr.,** *The Use of Proteolytic Enzymes for the Synthesis of Specific Peptide Bonds in Globular Proteins,* Offord, R. E. and DiBello, C., Eds., Academic Press, New York, 1978, 255.

43. **Finkenstadt, R. and Laskowski, M., Jr.,** Resynthesis by trypsin of the cleaved peptide bond in modified soybean trypsin inhibitor, *J. Biol. Chem.,* 242, 771, 1967.

44. **Leary, T. R. and Laskowski, M., Jr.,** Enzymatic replacement of Arg_{63} by Trp_{63} in the reactive site of soybean trypsin inhibitor (Kunitz) — an intentional change from tryptic to chymotryptic specificity, *Fed. Proc. Fed. Am. Soc. Exp. Biol.,* 32, 465, 1973.

45. **Jering, H. and Tschesche, H.,** Austausch von Lysin gegen Arginin, Phenylalanin und Tryptophan im reaktiven Zentrum des Trypsin-Kallikrein-Inhibitors (Kunitz), *Angew. Chem.,* 86, 704, 1974.

46. **Jering, H. and Tschesche, H.,** Replacement of lysine by arginine, phenylalanine, and tryptophan in the reactive site of the bovine trypsin-kallikrein inhibitor (Kunitz) and change of the inhibitory properties, *Eur. J. Biochem.,* 61, 453, 1976.

47. **Kassell, B., Radicevic, M., Ansfield, M. J., and Laskowski, M., Sr.,** The basic trypsin inhibitor of bovine pancreas. IV. The linear sequence of the 58 amino acids, *Biochem. Biophys. Res. Commun.,* 18, 255, 1965.

48. **Kassell, B. and Laskowski, M., Sr.,** The basic trypsin inhibitor of bovine pancreas. V. The disulfide linkages, *Biochem. Biophys. Res. Commun.,* 20, 463, 1965.

49. **Jering, H. and Tschesche, H.,** Darstellung des aktiven Derivates von Rinder-Trypsin-Kallikrein-Inhibitor (Kunitz) mit im aktiven Zentrum geöffneter Peptidbindung, *Angew. Chem.,* 86, 702, 1974.

50. **Jering, H. and Tschesche, H.,** Preparation and characterization of the active derivative of bovine trypsin-kallikrein inhibitor (Kunitz) with the reactive site lysine-15—alanine-16 hydrolyzed, *Eur. J. Biochem.,* 61, 443, 1976.

51. **Ardelt, W. and Laskowski, M., Jr.,** Thermodynamics and kinetics of the hydrolysis and resynthesis of the reactive site peptide bond in turkey ovomucoid third domain by aspergillopeptidase B, *Acta Biochim. Polonica,* 30, 115, 1983.

52. **Kato, J., Kohr, W. J., and Laskowski, M., Jr.,** Limited proteolysis of ovomucoids caused by staphylococcal proteinase, *Fed. Proc. Fed. Am. Soc. Exp. Biol.,* 36, 764, 1977.

53. **Wieczorek, M. and Laskowski, M., Jr.,** Formation of covalent hybrids from amino-terminal and carboxy-terminal fragments of two ovomucoid third domains, *Biochemistry,* 22, 2630, 1983.

54. **Richards, R. M. and Vithayathil, P. J.,** The preparation of subtilisin-modified ribonuclease and the separation of the peptide and protein components, *J. Biol. Chem.,* 234, 1459, 1959.

55. **Homandberg, G. A. and Laskowski, M. Jr.,** Enzymatic resynthesis of the hydrolyzed peptide bond(s) in ribonuclease S, *Biochemistry,* 18, 586, 1979.

56. **Homandberg, G. A., Komoriya, A., Juillerat, M., and Chaiken, I. M.,** Enzymatic conversion of selected noncovalent complexes of native or synthetic fragments to covalent forms, in *Peptides, Structure, and Biological Function,* Gross, E. and Meienhofer, J., Eds., Pierce Chem., Rockford, Ill., 1980, 597.

57. **Komoriya, A., Homandberg, G. A., and Chaiken, I. M.,** Enzymatic fragment condensation using kinetic traps, in *Peptides 1980, Proc. 16th Eur. Peptide Symp.,* Brunfeldt, K., Ed., Scriptor, Kopenhagen, 1981, 378.

58. **Homandberg, G. A., Komoriya, A., and Chaiken, I. M.,** Enzymatic condensation of nonassociated peptide fragments using a molecular trap, *Biochemistry,* 21, 3385, 1982.

59. **Mitchel, W. M. and Harrington, W. F.,** Purification and properties of a clostridiopeptidase B (clostripain) *J. Biol. Chem.,* 243, 4683, 1968.

60. **Homandberg, G. A. and Chaiken, I. M.,** Trypsin-catalyzed conversion of staphylococcal nuclease-T fragment complexes to covalent forms, *J. Biol. Chem.,* 255, 4903, 1980.

61. **Taniuchi, H., Anfinsen, C. B., and Sodja, A.,** Nuclease-T: an active derivative of staphylococcal nuclease composed of two noncovalently bonded peptide fragments, *Proc. Natl. Acad. Sci. U.S.A.,* 58, 1235, 1967.

62. **Taniuchi, H. and Anfinsen, C. B.,** Steps in the formation of active derivatives of staphylococcal nuclease during trypsin digestion, *J. Biol. Chem.,* 243, 4778, 1968.

63. **Komoriya, A., Homandberg, G. A., and Chaiken, I. M.,** Enzyme-catalyzed formation of semisynthetic staphylococcal nuclease using a new synthetic fragment, (48-Glycine) synthetic-(6-49), *Int. J. Peptide Protein, Res.*, 16, 433, 1980.

64. **Tesser, G. I. and Boon, P. J.,** Semisynthesis in protein chemistry, *Recl. Trav. Chim. Pays-Bas,* 99, 289, 1980.

65. **Margoliash, E., Smith, E. L., Kreil, G., and Tuppy, H.,** Amino-acid sequence of horse heart cytochrome c. The complete amino-acid sequence, *Nature (London),* 192, 1125, 1961.

66. **Harris, D. E. and Offord, R. E.,** A functioning complex between tryptic fragments of cytochrome c. A route to the production of semisynthetic analogues, *Biochem. J.,* 161, 21, 1977.

67. **Westerhuis, L. W., Tesser, G. I., and Nivard, R. J. F.,** Enzymatic synthesis of a peptide bond between a tryptic fragment of horse heart cytochrome c and a synthetic model peptide, *Recl. Trav. Chim. Pays-Bas,* 99, 400, 1980.

68. **Juillerat, M. and Homandberg, G. A.,** Clostripain-catalyzed re-formation of a peptide bond in a cyto chrome c fragment complex, *Int. J. Peptide Protein Res.,* 18, 335, 1981.

69. **Proudfoot, A. E. I. and Wallace, C. J. A.,** Cytochrome c semisynthesis using enzymic resynthesi techniques, in *Peptides 1982, Proc. 17th Eur. Peptide Symp.,* Blahá, K. and Maloň, P., Eds., Walter d Gruyter, Berlin, 1983, 353.

70. **Graf, L. and Li, C. H.,** Human somatotropin: covalent reconstitution of two polypeptide contiguou fragments with thrombin, *Proc. Natl. Acad. Sci. U.S.A.,* 78, 6135, 1981.

71. **Nagai, K., Enoki, Y., Tomita, S., and Teshima, T.,** Trypsin-catalyzed synthesis of peptide bond i human hemoglobin. Oxygen binding characteristics of Gly-NH$_2$(142α)Hb, *J. Biol. Chem.,* 257, 1622, 1982

72. **O'Donnell, S., Mandaro, R., Schuster, T. M., and Arnone, A.,** X-ray diffraction and solution studie of specifically carbamylated human hemoglobin A. Evidence for the location of a proton- and oxygen linked chloride binding site at valine 1 α, *J. Biol. Chem.,* 254, 12204, 1979.

Chapter 11

KINETICS OF PROTEASE-CONTROLLED SYNTHESES OF PEPTIDE BONDS

The rational application of proteases in preparative peptide synthesis requires an insight into specificity and efficiency of these enzymes as catalysts for peptide bond formation. These properties of the proteases are most reliably described by kinetic parameters such as the Michaelis constant, K_m, (substrate concentration at half-maximal velocity) and the catalytic constant, k_{cat} or k_o (maximal initial velocity per unit enzyme concentration).[1] However, though the kinetics of the protease-catalyzed hydrolysis of peptide-, ester-, and amide bonds have been extensively investigated, reports of kinetic studies dealing with the protease-catalyzed synthesis of peptide bonds are relatively sparse. Even if the same structural features of the substrates participate in both the mechanism of peptide bond synthesis and hydrolysis, the kinetic constants for hydrolysis are only conditionally applicable to determining the ability of a given protease to catalyze the peptide-bond-forming step. Indeed, the boundary conditions for the latter process will normally differ markedly from those permitting peptide-bond hydrolysis, since they are optimized to favor peptide bond synthesis while suppressing the hydrolytic process. Therefore, further exploration of the kinetic aspects of the peptide synthetic potential of the proteinases could lead us toward a more intelligent use of these enzymes in preparative peptide synthesis.

Gawron et al.[2] carried out a kinetic analysis of the first well-defined chymotryptic peptide bond synthesis performed by Bergmann and Fruton.[3] The initial reaction velocities of the synthesis of Bz-Tyr-Gly-NHPh from Bz-Tyr-OH and H-Gly-NHPh indicated both a Michaelis-Menten relationship between the rate of product formation and the concentration of the first substrate, Bz-Tyr-OH, and a linear dependency of the rate of product formation on the concentration of the second substrate, H-Gly-NHPh. The Michaelis constant, K_m, of Bz-Tyr-OH was determined to be 80 mM and the maximal velocity of synthesis was given at 25 mM/sec (per mole of chymotrypsin and H-Gly-NHPh). The authors concluded from these results, that the molecular mechanism of the protease-catalyzed synthesis was the reverse of the hydrolytic reaction.

The balance between aminolytic and hydrolytic cleavage of acyl-enzyme complexes of the form N^α-Ac-(Gly)$_n$-Phe(NO$_2$)-chymotrypsin was kinetically analyzed by Petkov and Stoineva.[4] The ratio of rate constants of aminolysis and hydrolysis, the so-called partitioning ratio, provided a quantitative measure of the "true" nucleophile reactivity; i.e., it describes the capacity of a given amine component to reduce the extent of hydrolysis. The kinetic data obtained strongly emphasized the influence of the P'_2-site of the amine components on chymotryptic peptide synthesis. For example, those amine components having a hydrophobic amino acid residue in the P'_2-position are significantly more efficient than those having a charged or polar amino acid unit in this position. Chymotryptic esterase activities were favored with growing chain length, i.e., with increasing number, n, of glycine residues in the acyl-group donor N^α-Ac-(Gly)$_n$-Phe(NO$_2$)-OMe, while with the same condition the reactivity of the nucleophile acceptors simultaneously decreased.

The molecular mechanism of α-chymotryptic and β-tryptic peptide bond formation has been studied by Riechmann and Kasche.[5,6] Using Ac-Tyr-OEt, Ac-Tyr-pNA, and Bz-Arg-pNA as acyl-group donors and various amino acid amides and esters as acyl-group acceptors (nucleophiles), the partition of the resulting Ac-Tyr-chymotrypsin, Ac-Tyr-trypsin, and Bz-Arg-trypsin complex, respectively, between aminolysis and hydrolysis was investigated. The tendency toward aminolytic deacylation of the complexes increased with growing hydrophobicity of the nucleophiles. From the initial rate data the authors propose a mechanism which involves two acyl-enzyme complexes: one with and one without a noncovalently bound nucleophile moiety. Both complexes can be cleaved hydrolytically, however, aminolytic cleavage can only proceed from the complex in which the nucleophile has bound prior to deacylation.

A systematic study on the kinetics of thermolysin-catalyzed peptide bond formation was reported by Wayne and Fruton.[7] The reaction rates determined for the coupling of various N^α-substituted amino acids and peptides with leucine anilide revealed that thermolysin shows a strong preference for hydrophobic amino acids in the P_1-site. Furthermore, hydrophobic L- although not D-amino acid residues in the P_2-position tend to enhance the progress of synthesis. The influence of amine components on the reaction velocity was studied by means of a variety of syntheses in which Z-Phe-OH was coupled with different amine components. The comparison of the resulting kinetic data revealed that the Michaelis constants, $K_{m,\ Z\text{-Phe-OH}}$, and the catalytic constants, k_o, were strongly dependent upon the nature of the respective amine component. The numerical values of $K_{m,\ K\text{-Phe-OH}}$ ranged from 7.8 mM (H-Leu-NHPh) to 62.8 (H-Phe-Gly-OMe) while those of k_o ranged from 1.45/sec (H-Phe-OEt) to 192/sec (H-Phe-Gly-OMe). Since the k_o values measured in the presence of different amine components varied more than the K_m values, the authors concluded that the association of the amine component with thermolysin is accompanied by conformational changes at the binding site for the substrate Z-Phe-OH, i.e., that a substrate synergism governs the reaction. The initial rates observed for the synthesis of Z-Phe-Phe-OMe from Z-Phe-OH as the variable substrate in the presence of different fixed concentrations of H-Phe-OMe, were consistent with the kinetics of a "random bireactant" system. The numerical value of $K_{m,\ Z\text{-Phe-OH}}$ amounted to 9.7 mM, whereas the $K_{m, H\text{-Phe-OMe}}$ was approximately 300 mM. The magnitude of the estimated $K_{m,\ Phe\text{-OMe}}$ also explained why Oyama et al.[8] could not determine the K_m value for H-Phe-OMe when they studied the kinetics of the thermolysin-mediated synthesis of the aspartame precursor Z-Asp-Phe-OMe. The authors therefore assumed that the velocity of the enzymatic coupling of Z-Asp-OH and H-Phe-OMe was a linear function of the concentration of phenylalanine methyl ester. On the other hand, when the rate of product formation was plotted as a function of the Z-Asp-OH concentration, the typical Michaelis-Menten behavior was observed; the K_m value for Z-Asp-OH was determined to be 10.3 mM.

The specificity of pepsin-controlled peptide syntheses was studied by Bozler et al.[9] by exploring the influence of a series of carboxyl- and amine components on the initial rates of the respective synthetic processes. Comparative analyses of the observed kinetic data revealed that the synthetic specificity largely paralleled the hydrolytic specificity. Thus it was shown that pepsin which demonstrates a primary specificity for phenylalanine residues occupying the P_1-site of a given substrate during proteolytic reactions, shows a similar substrate specificity for proteosynthetic reactions.[10] In addition, pepsin exhibits a preference for hydrophobic residues in the P_2-position, and as such a phenylalanine-anilide in the P_1'-site was superior to a leucine anilide in this position. From the point of view of peptic peptide synthesis these results underline — apart from the predominant preference of the protease for the Phe-Phe-bond — the importance of secondary substrate-enzyme interactions for peptide synthetic purposes. In this context it should be added that partially-acetylated pepsin, which exhibits enhanced proteolytic activities, as compared to native pepsin, also catalyzes peptide bond formation more efficiently.[9]

A direct comparison of the rate constants of the forward- and reverse reaction of protease-catalyzed processes was enabled by kinetic studies on protease inhibitors. The ratio of the catalytic constants of synthesis and hydrolysis $k_{o,syn}/k_{o,hyd}$ was often found to be greater than unity.[11] For example, the aspergillopeptidase B-catalyzed formation of the peptide bond at the active site of the third domain of ovomucoid protein (*vide supra,* Chapter 10, Section III) proceeds 7.7×10^3 times faster than its hydrolysis as catalyzed by the same protease.[11] Tryptic resynthesis of the active site peptide bond in the trypsin-kallikrein inhibitor (TKI) and in the trypsin inhibitor (STI) takes place at, respectively, a 90-[12] and 5 times[13] higher rate than the corresponding hydrolyses. The above results are hardly in agreement with the general view that hydrolysis largely predominates over synthesis in a protease-catalyzed process. However, this apparent contradiction is readily explained if one considers the rigid molecular environment of the peptide bonds in question. Upon cleavage of these bonds, only a negligible entropy gain accrues and the newly formed hydrolysis products may hardly interact more favorably in energy terms with the solvents than the intact molecule. Beyond that, the zwitterionic character of the intact inhibitor and of the modified, split inhibitor differs significantly from the charged nature of small model peptides or amino acids, the acidity or basicity of which is most commonly used to estimate the respective prospects of peptide bond formation and cleavage. A study on the trypsin inhibitor by Mattis and Laskowski, Jr.[14] sheds some light upon the pronounced differences between the acidic and basic properties of free amino acids and the corresponding residues in proteins. The active site peptide bond of the inhibitor is that between the Arg_{63}-Ile_{64} linkage (cf. Chapter 19, Sect. III, Figure 3). Within the modified inhibitor (STI*) in which this bond is cleaved the pK_a-value of the carboxyl group of the arginine$_{63}$ residue is 3.56 while the pK value of the amino group of the isoleucine$_{64}$ residue is 7.89. In contrast, the carboxyl function of free arginine has a pK_a of about 2.17 and the amino function of isoleucine has a pK_a about 9.68.

Systematic studies dealing with the kinetics of trypsin-catalyzed conversion of porcine to human insulin on an industrial scale have been reported recently by Markussen and Vølund.[15,16] Esterified human insulin was obtained either from porcine insulin by an *in situ* exchange of the original AlaB$_{30}$-residue for a threonine ester, i.e., via transpeptidation, or from des-(Ala B$_{30}$)-porcine insulin (DAI), to which the respective threonine derivative was coupled. Michaelis-Menten enzyme kinetics were applicable to the trypsin-mediated semi-syntheses of human insulins, these being performed in predominantly organic media with a water content of 20% (w/v). However, K_m values could not be determined as saturation of the enzyme with substrate was not achieved even though the insulin concentration was as high as 8 mM. From the intercept with the X-axis and the slope of a Lineweaver-Burk plot,[17] the authors estimated the K_m- and the V_{max} value for the coupling of H-Thr-OMe to DAI in the presence of 0.2 mM trypsin to be greater than 0.1 M and 18 mM/hr, respectively. The extremely high K_m value was ascribed to the fact that the enzyme-substrate system existed largely in the dissociated state as a consequence of the presence of the organic co-solvents. For similar reasons K_m values could not be determined for the tryptic conversion of native porcine insulin into human insulin esters. The initial velocities of the reaction between DAI and H-Thr-OMe or H-Thr(Bu^t)-OBu^t were found to be, respectively, 80 and 15 times greater than those for the corresponding tryptic semisyntheses of human insulin esters from porcine insulin via direct AlaB$_{30}$ → Thr exchange. The initial rate for the coupling of DAI with H-Thr-OMe exceeded that observed for H-Thr(Bu^t)-OBu^t, the apparent first-order rate constants — normalized to 1 mM trypsin — being 75/hr and 27/hr — respectively. Conversely, the initial rate of direct conversion of porcine insulin into human insulin ester was higher in the presence of H-Thr(Bu^t)-OBu^t than of H-Thr-OMe. Here, the apparent first-order rate constants were 0.93/hr and 1.84/hr, respectively. The release of the C-terminal alanine residue from the B-chain of porcine insulin was obviously inhibited by H-Thr-OMe. According to the authors, this inhibitory effect can be attributed to the ability of H-Thr-OMe to exclude water from the acyl-enzyme complex. The yields of human insulin esters ranged from 90 to 96% after 24 hr at 12°C. The threonine derivatives were added in a large molar excess — 1 M against 2 to 8 mM of DAI or porcine insulin. The incubation medium having pH 4.5 was composed of about 60% dimethylacetamide, 20% water, and 2.5 M acetic acid.

Our own studies in this area have focused on the kinetic aspects of the previously mentioned enzymatic syntheses of the enkephalins;[18,19] (see also Chapter 8, Section II) with the aim of gaining further insights into the proteosynthetic potential of various proteases. These investigations particularly aimed at the numerical evaluation of kinetic constants in order to provide a measure of the specificity and catalytic efficiency of papain- and α-chymotrypsin

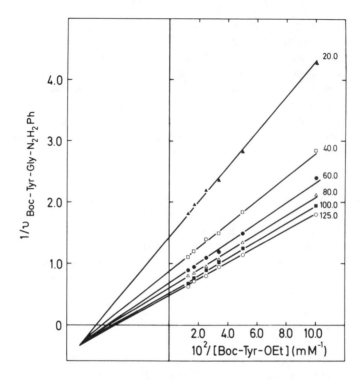

FIGURE 1. Kinetics of Boc-Tyr-Gly-N$_2$H$_2$Ph synthesis from Boc-Tyr-OEt and H-Gly-N$_2$H$_2$Ph via chymotrypsin catalysis. The initial velocity (mM/sec) pattern of peptide bond formation with Boc-Tyr-OEt as the varied-concentration substrate is shown. The millimolar concentrations of H-Gly-N$_2$H$_2$Ph were held constant at the values indicated at the ends of the lines.

in peptide bond formation.[20] Furthermore, we hoped to develop a concept of the molecular mechanism of protease-catalyzed peptide bond synthesis from observations of the initial rates of reactions.

Kinetic analysis of the chymotrypsin-mediated synthesis of the dipeptide Boc-Tyr-Gly-N$_2$H$_2$Ph revealed the following picture: the initial-velocity pattern, obtained from a series of dipeptide syntheses using Boc-Tyr-OEt as the variable substrate in the presence of different fixed concentrations of H-Gly-N$_2$H$_2$Ph, showed a series of straight intersecting lines in a conventional double-reciprocal Lineweaver-Burk plot (Figure 1).[17] A similar pattern emerged when the varied-concentration and fixed-concentration substrates were interchanged.

The family of converging double-reciprocal plots fits a Michaelis-Menten rate law for two-substrate reaction,

$$v = \frac{V_{syn}[A][B]}{K_s^A K_m^B + K_m^A[B] + K_m^B[A] + [A][B]} \tag{1}$$

where v is the initial velocity for the formation of Boc-Tyr-Gly-N$_2$H$_2$Ph, V_{syn} is the maximum velocity of synthesis, [A] and [B] are, respectively, the concentrations of the first and the second substrate, K_m^A and K_m^B represent their Michaelis constants, and finally, K_s^A may be the dissociation (or other) constant for the complex between substrate A and the enzyme.

The observed intersecting initial-rate pattern appears to be diagnostic of a sequential mechanism where both substrates bind to the enzyme before the release of either product. However, when the initial velocity of ethanol liberation was measured as a function of varied

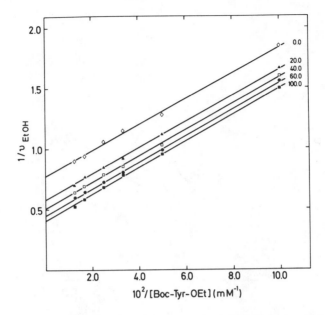

FIGURE 2. Kinetics of chymotrypsin-catalyzed ethanol release. Initial velocities (mM/sec) of ethanol release are plotted in double-reciprocal form with Boc-Tyr-OEt as the variable substrate. The millimolar concentrations of H-Gly-N$_2$H$_2$Ph were fixed at the values indicated at the ends of the lines.

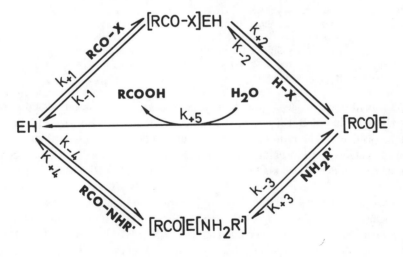

FIGURE 3. Reaction scheme for concurrent peptide bond formation and hydrolysis as catalyzed by α-chymotrypsin.

Boc-Tyr-OEt concentrations at different fixed concentrations of H-Gly-N$_2$H$_2$Ph, a set of parallel lines was obtained in a double-reciprocal plot (Figure 2).

These results are compatible neither with a sequential mechanism nor with a simple ping-pong mechanism in which the first product is released before the second substrate is bound to the enzyme. However, the kinetic dates are consistent with a ping-pong mechanism modified by a hydrolytic branch, as summarized in Figure 3. Analogous mechanisms involving covalent enzyme-substrate complexes have been extensively evidenced for serine- and cysteine proteases.[21]

To determine the order of substrate addition, the initial-velocity ratio of peptide bond

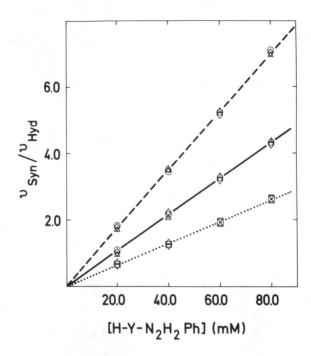

FIGURE 4.   Plots of peptide synthesis versus hydrolysis in the
presence of α-chymotrypsin. The initial velocity ratios of peptide
synthesis to hydrolysis (v_{syn}/v_{hyd}) are plotted against the concentrations of the acyl acceptors (H-Y-N$_2$H$_2$Ph). The graphs refer to the
respective substrates as follows; (·······), Boc-Tyr-OEt and H-Gly-
N$_2$H$_2$Ph;   (————),   Boc-Gly-Phe-OEt and H-Leu-N$_2$H$_2$Ph;
(- - - - - -), Boc-Gly-Phe-OEt and H-Met-N$_2$H$_2$Ph. Concentrations of
the acyl donors: △, 20.0 m*M*; □, *40.0 mM;* ○, 50.0 m*M;* ▽, 80.0
m*M,* ◇, 90.0 m*M.*

formation to hydrolysis v_{syn}/v_{hyd}, was plotted both as a function of various concentrations
of H-Gly-N$_2$H$_2$Ph and of Boc-Tyr-OEt at fixed concentrations of Boc-Tyr-OEt and H-Gly-
N$_2$H$_2$Ph, respectively. The ratio v_{syn}/v_{hyd} was found to be directly proportional to the concentration of H-Gly-N$_2$H$_2$Ph (Figure 4) and independent of the Boc-Tyr-OEt concentration.

According to Frère,[22] these kinetic features are compatible with an ordered mechanism
in which Boc-Tyr-OEt binding first and H-Gly-N$_2$H$_2$Ph binds second to α-chymotrypsin.
The proposed mechanism for the α-chymotrypsin-mediated reaction, as outlined in Figure
3, is characterized by the binding of the first substrate RCO-X to the proteinase EH thus
forming a binary enzyme-acyl-donor complex [RCO-X]-EH. The first product H-X is subsequently released, leaving a covalent acyl-enzyme complex, [RCO]-E, which can then
transfer an acyl group either to the second substrate the amine, NH$_2$R, i.e., aminolysis, to
give the second product the peptide RCO-NHR, or to water, i.e., hydrolysis, to give the
alternative second product RCO$_2$H. Thus to illustrate the chymotrypsin-mediated synthesis
of Boc-Tyr-Gly-N$_2$H$_2$Ph, the terms RCO-X, NH$_2$R, H-X, RCO-NHR, and RCO$_2$H given in
the proposed reaction pathway can be replaced by Boc-Tyr-OEt, H-Gly-N$_2$H$_2$Ph, ethanol,
Boc-Tyr-Gly-N$_2$H$_2$Ph, and Boc-Tyr-OH, respectively.

On the basis of a ping-pong mechanism modified by a hydrolytic shunt (Figure 3), the
initial rate law for the synthesis of Boc-Tyr-Gly-N$_2$H$_2$Ph is given by Equation 1 whereas
the initial-velocity of the ethanol release, and of the appearance of Boc-Tyr-OH, are described
by the following Equations 2 and 3, respectively,[23]

$$\upsilon = \frac{V_{hyd}[A]}{K_s^A + [A](1 + [B]/K_m^B)(1 + [B]/K_i^B)^{-1}} \qquad (2)$$

$$\upsilon = \frac{V_{hyd}[A]}{K_s^A(1 + [B]/K_i^B) + [A](1 + [B]/K_m^B)} \qquad (3)$$

where V_{hyd} is the maximum velocity of hydrolysis, K_s^A represents the Michaelis constant for Boc-Tyr-OEt in the hydrolysis reaction, i.e., in the absence of H-Gly-N_2H_2Ph. The kinetic constant K_i^B is composed of a combination of rate constants that relate to the reaction of the covalent acyl-enzyme intermediate, [RCO]-E, with water and the second substrate, namely, H-Gly-N_2H_2Ph. The pattern of parallel lines obtained from the reciprocal plot of the initial rate of ethanol liberation in the presence of different concentrations of H-Gly-N_2H_2Ph (Figure 2) suggests uncompetitive activation of ethanol release by H-Gly-N_2H_2Ph. Activation rather than inhibition is indicated because $K_m^B > K_i^B$.[24] Initial-velocity patterns similar to those described above, were also obtained for the α-chymotrypsin-catalyzed syntheses of Boc-Gly-Phe-Leu-N_2H_2Ph and Boc-Gly-Phe-Met-N_2H_2Ph starting from the acyl donor Boc-Gly-Phe-OEt and the acyl acceptor H-Leu-N_2H_2Ph and H-Met-N_2H_2Ph, respectively. The values for the respective kinetic parameters are given in Table 1.

From the mechanism shown in Figure 3, it follows that the same acyl-enzyme complex is involved in both aminolysis and hydrolysis. Given the aims of peptide synthetic chemistry, it is desirable to enhance the aminolytic deacylation of this complex during peptide synthesis at the expense of its hydrolytic cleavage. As illustrated in Figure 4, the initial-velocity ratios of peptide bond formation to hydrolysis are linearly dependent on the concentration of the individual acyl acceptor, but are independent of the acyl-donor concentration (results not shown). Consequently, the proteinase-controlled reactions can be shifted in favor of peptide synthesis by increasing the concentration of the acyl acceptor. The kinetic parameters of Equations 1 to 3 are defined in terms of rate constants (cf. Figure 3) as follows:

$$K_s^A = \frac{k_{+5}(k_{-1} + k_{+2})}{k_{+1}(k_{+2} + k_{+5})} \qquad (4)$$

$$K_m^A = \frac{k_{+4}(k_{-1} + k_{+2})}{k_{+1}(k_{+2} + k_{+4})} \qquad (5)$$

$$K_m^B = \frac{(k_{-3} + k_{+4})(k_{+2} + k_{+5})}{k_{+3}(k_{+2} + k_{+4})} \qquad (6)$$

$$K_i^B = \frac{k_{+5}(k_{-3} + k_{+4})}{k_{+3}k_{+4}} \qquad (7)$$

$$V_{syn} = \frac{k_{+2}k_{+4}[E]_o}{k_{+2} + k_{+4}} \qquad (8)$$

$$V_{hyd} = \frac{k_{+2}k_{+5}[E]_o}{k_{+2} + k_{+5}} \qquad (9)$$

In the following paragraph, a kinetic analysis of the papain-mediated peptide-bond-forming reactions carried out during the enkephalin syntheses is presented.[20]

When Boc-Gly-OH and H-Phe-N_2H_2Ph were used as substrates and the initial rate of

Table 1
KINETIC CONSTANTS FOR α-CHYMOTRYPSIN-CATALYZED PEPTIDE SYNTHESIS

Kinetic constants	Boc-Tyr-OEt H-Gly-N₂H₂Ph	Boc-Gly-Phe-OEt	
		H-Leu-N₂H₂Ph	H-Met-N₂H₂Ph
K_s^A(mM)	14.6 ± 0.6	12.5 ± 1.0	11.3 ± 1.2
	13.8 ± 0.8	11.5 ± 1.1	12.7 ± 1.3
K_m^A(mM)	38.2 ± 1.4	47.3 ± 3.8	52.0 ± 5.6
K_m^B(mM)	81.2 ± 2.3	63.5 ± 5.9	56.4 ± 5.0
K_i^B(mM)	27.5 ± 4.0	17.5 ± 2.0	12.8 ± 1.2
	30.2 ± 3.5	18.4 ± 2.3	12.0 ± 1.5
V_{syn}(mM/sec)	3.48 ± 0.30	2.10 ± 0.21	2.45 ± 0.25
V_{hyd}(mM/sec)	1.30 ± 0.10	0.61 ± 0.08	0.52 ± 0.08
$k_{o,syn}$(sec⁻¹)	26.0 ± 2.3	21.0 ± 2.0	24.5 ± 2.4
$k_{o,hyd}$(sec⁻¹)	9.7 + 0.7	6.1 ± 0.8	5.2 + 0.7

Note: The values are assigned on the basis of a "ping-pong" mechanism modified by a hydrolytic branch as depicted in Figure 3. (For a more detailed description of the numerical evaluation of kinetic constants cf. Reference 20.)

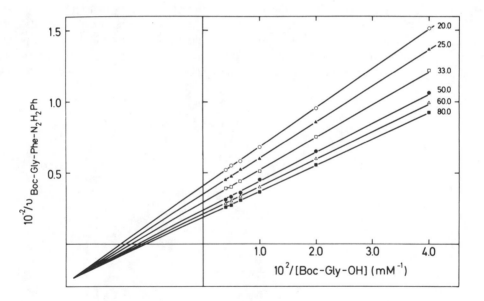

FIGURE 5.   Kinetics of papain-catalyzed synthesis of Boc-Gly-Phe-N₂H₂Ph from Boc-Gly-OH and H-Phe-N₂H₂Ph. The initial rate (mM/sec) pattern of peptide synthesis with Boc-Gly-OH as the variable substrate is shown. The millimolar concentrations of H-Phe-N₂H₂Ph were fixed at the values indicated at the ends of the lines.

papain-controlled synthesis of Boc-Gly-Phe-N₂H₂Ph was plotted against the concentration of Boc-Gly-OH (Figure 5) and of H-Phe-N₂H₂Ph (results not shown), intersecting initial-velocity patterns were obtained in conventional double-reciprocal plots. Here the observed kinetic features, which are generally diagnostic of sequential reaction pathways, are also compatible with the mechanism of condensation reactions, such as peptide bond formation, that generally appear to be ping-pong Bi-Bi with water as the first product, but which nevertheless give apparent sequential Bi-Uni kinetics. (Bi-Bi designates a chemical reaction in which two products are formed from two educts, while Bi-Uni represents a reaction in which two educts are condensed to form one product.) To further characterize the kinetic mechanism of the papain-mediated reaction and in particular to differentiate between random and ordered mechanisms, initial rate measurements in the presence of substrate analogues acting as alternative substrates were performed.[25] These studies revealed, that H-Leu-N$_2$H$_2$Ph acted competitively with respect to H-Phe-N$_2$H$_2$Ph and noncompetitively with respect to Boc-Gly-OH as an inhibitor of papain-catalyzed synthesis of Boc-Gly-Phe-N$_2$H$_2$Ph. Similarly, Boc-Gln-OH functioned as a competitive inhibitor with respect to Boc-Gly-OH. However, while the inhibition patterns observed with H-Leu-N$_2$H$_2$Ph as the alternative substrate for H-Phe-N$_2$H$_2$Ph during the first set of studies showed a linear dependence on the concentration of the substrate analogue H-Leu-N$_2$H$_2$Ph, the alternative substrate Boc-Gln-OH acted as a nonlinear, hyperbolic inhibitor with respect to H-Phe-N$_2$H$_2$Ph. These results are inconsistent both with a random pathway and with an ordered pathway in which H-Phe-N$_2$H$_2$Ph is the first substrate to bind to papain.

In fact the kinetic data presented above define the papain-catalysed peptide bond formation as a sequential ordered reaction, in which Boc-Gly-OH binds to the proteinase before H-Phe-N$_2$H$_2$Ph. This is illustrated in the scheme depicted in the outer rim of schemes (Figures 6A and B), when RCO-X, H-X, NH$_2$-R and RCO-NHR may be replaced by Boc-Gly-OH, H$_2$O, H-Phe-N$_2$H$_2$Ph and Boc-Gly-Phe-N$_2$H$_2$Ph, respectively.

The initial-velocity of Boc-Gly-Phe-N$_2$H$_2$Ph synthesis is given by the rate of Equation 1 (see above), the kinetic constants of which and those of Equations 14 and 15, (see below) are related in the following manner to the constants given in Figure 6 (analogous definitions hold for the "primed" constants):

$$K_s^A = \frac{k_{-1}}{k_{+1}} \tag{10}$$

$$K_m^A = \frac{k_{+4}(k_{-1} + k_{+2})}{k_{+1}(k_{+2} + k_{+4})} \tag{11}$$

$$K_m^B = \frac{k_{+2}(k_{-3} + k_{+4})}{k_{+3}(k_{+2} + k_{+4})} \tag{12}$$

$$V_{syn} = \frac{k_{+2}k_{+4}[E]_o}{k_{+2} + k_{+4}} \tag{13}$$

The modified rate expression for this mechanisms, when the alternative substrate, Boc-Gln-OH, is used alone with the common substrate, Boc-Gly-OH, is

$$v = \frac{V_{syn}[A][B]}{K_m^A[B](1 + [A']/K_i) + K_m^B[A] + K_s^A K_m^B(1 + [A']/K_i) + [A][B]} \tag{14}$$

where [A'] is the concentration of the alternative substrate, and $K_i = (K_m'^B + [B]/(K_s'^{A'} K_m^{A'} + K_m^{A'} [B]))$[26] with $K_m^{A'}$, $K_s^{A'}$ and $K_m'^B$ being the K_m for Boc-Gln-OH, the dissociation constant of Boc-Gln-OH from the enzyme complex and the K_m for H-Phe-N$_2$H$_2$Ph, respectively, when Boc-Gln-OH is the co-substrate. The mechanism of the modified reaction is shown in Figure 6A, where R"CO-X represents Boc-Gln-OH.

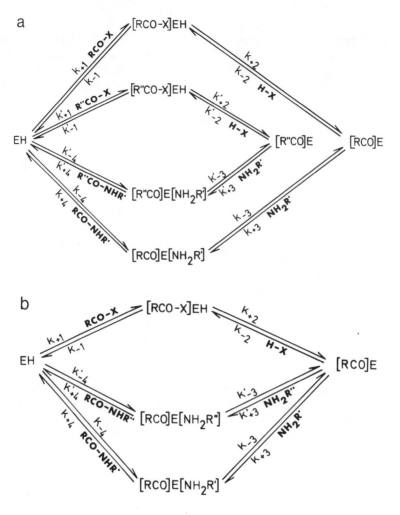

FIGURE 6.   Reaction scheme for papain-catalyzed peptide synthesis in the presence of an alternative substrate for (A) the first substrate and (B) the second substrate.

When the alternative substrate H-Leu-N_2H_2Ph is used along with H-Phe-N_2H_2Ph, the modified rate law will be (re-arranged in accordance with Segel[27]):

$$v = \frac{V_{syn}[A][B]}{K_m^A[B] + K_m^B[A](1 + [B']/K_m^{B'}) + K_s^A K_m^B(1 + [B']/K_{ii}) + [A][B]} \quad (15)$$

where [B'] represents the concentration of H-Leu-N_2H_2Ph and $K_{ii} = K_s^A K_m^{B'}/K_m^{'A}$. Here $K_m^{'A}$ denotes the K_m for Boc-Gly-OH in the presence of the alternative substrate, H-Leu-N_2H_2Ph, and $K_m^{B'}$ is the K_m for H-Leu-N_2H_2Ph in place of H-Phe-N_2H_2Ph. The pathway of this modified reaction is outlined in Figure 6B, where NH_2R'' represents H-Leu-N_2H_2Ph.

Kinetic studies on the papain-catalysed formation of Boc-Tyr-(Bzl)-Gly-Gly-Phe-Leu-N_2H_2Ph and of the corresponding methionine derivative from the acyl donor Boc-Tyr(Bzl)-Gly-OH and the acyl acceptors H-Gly-Phe-Leu-N_2H_2Ph and H-Gly-Phe-Met-N_2H_2Ph, respectively, show that the reaction scheme (Figure 6) proposed for the synthesis of Boc-Gly-Phe-N_2H_2Ph is also valid for the enzymatic preparation of the above pentapeptides. (The numerical values of the kinetic parameters for the papain-catalyzed syntheses are given in Table 2).

Table 2
KINETIC CONSTANTS FOR PAPAIN-CATALYZED PEPTIDE SYNTHESIS

	Substrates		
		Boc-Tyr(Bzl)Gly-OH H-Gly-Phe-X-N₂H₂Ph	
Kinetic constants	Boc-Gly-OH H-Phe-N₂H₂Ph	Leu	Met
K_s^A (mM)	43.0 ± 4.1	4.8 ± 0.5	4.5 ± 0.5
K_m^A (mM)	135 ± 14	52.5 ± 3.8	50.8 ± 4.3
K_m^B (mM)	51.2 ± 4.2	65.0 ± 7.5	67.5 ± 6.4
V_{syn} (mM/sec)	0.09 ± 0.01	0.29 ± 0.03	0.30 ± 0.03
k_o (sec^{-1})	0.27 ± 0.03	3.16 ± 0.33	3.26 ± 0.33

	Substrates		Substrates
Kinetic constants	Boc-Gln-OH H-Phe-N₂H₂Ph	Kinetic constants	Boc-Gly-OH H-Leu-N₂H₂Ph
$K_s^{A'}$ (mM)	11.5 ± 1.3	$K_s^{A'}$ (mM)	40.8 ± 3.2
$K_m^{A'}$ (mM)	108 ± 12	$K_m^{A'}$ (mM)	127 ± 13
	113 ± 18		131 ± 17
$K_m^{B'}$ (mM)	50.5 ± 4.8	$K_m^{B'}$ (mM)	44.2 ± 3.7
			46.3 ± 4.2
$K_s^{A'} K_m^{B'}$ (mM)^2	575 ± 84	V_{syn} (mM/sec)	0.08 + 0.01
	610 ± 95		
V_{syn} (mM/sec)	0.01 ± 0.01	k_o (sec^{-1})	0.25 ± 0.03
k_o (sec^{-1})	0.30 ± 0.03		

Note: The values are assigned on the basis of sequential ordered mechanisms as shown in Figures 6A and B. In Part A, the kinetic data were determined from initial velocity measurements in the absence of alternative substrates. In Part B the values for the "primed" constants, which refer to alternative substrates, were obtained through initial rate studies in the presence of alternative substrates. (For a more detailed description of the numerical evaluation of kinetic constants cf. Reference 20.)

The catalytic potential of papain in peptide bond formation, which is reflected by the observed values for the individual catalytic constants, k_o, is significantly inferior to that found for α-chymotrypsin. This difference may be explained by the fact that in the contrast with the chymotrypsin-catalysis, the acyl donors participating in the papain-mediated reactions were unesterified. As mentioned previously, ester donors are utilized more efficiently in papain-controlled- as well as in chymotrypsin-controlled syntheses than are their corresponding free acids.

Thus in the present study, the catalytic constant for the papain-mediated synthesis of Boc-Gly-Phe-N₂H₂Ph could be increased by roughly one order of magnitude by replacing Boc-Gly-OH by Boc-Gly-OEt under otherwise comparable conditions. Consequently, for peptide synthesis via papain catalysis esterified acyl donors should be preferable to acyl donors having a free α-carboxyl group.

Surprisingly while preparing the enkephalins, we have found that the papain-catalyzed synthesis of simple dipeptides required an increased enzyme concentrations and prolonged incubation times relative to the synthesis of tetrapeptides and pentapeptides.[19] We therefore ascribe the pronounced differences between the catalytic constants, k_o, (Table 2) for the synthesis via papain catalysis of the dipeptides and of the pentapeptides to a more favorable interaction of the enlarged substrates with the extended active site of papain.[28]

REFERENCES

1. **Fruton, J. S.,** Proteinase-catalyzed synthesis of peptide bonds, *Adv. Enzymol. Relat. Areas Mol. Biol.,* 53, 239, 1982.
2. **Gawron, O., Glaid, A. J., III, Boyle, R. E., and Odstrchel, G.,** Kinetics of the chymotrypsin-catalyzed condensation of N-Benzoyl-L-tyrosine with glycylanilide, *Arch. Biochem. Biophys.,* 95, 203, 1961.
3. **Bergmann, M. and Fruton, J. S.,** Some synthetic and hydrolytic experiments with chymotrypsin, *J. Biol. Chem.,* 124, 312, 1938.
4. **Petkov, D. D. and Stoineva, I. B.,** Nucleophile specificity in chymotrypsin peptide synthesis, *Biochem. Biophys. Res. Commun.,* 118, 317, 1984.
5. **Riechmann, L. and Kasche, V.,** Kinetic studies on the mechanism and specificity of peptide synthesis catalyzed by the serine proteases α-chymotrypsin and β-trypsin, *Biochem. Biophys. Res. Commun.,* 120, 686, 1984.
6. **Riechmann, L. and Kasche, V.,** Peptide synthesis catalyzed by the serine proteinases chymotrypsin and trypsin, *Biochim. Biophys. Acta,* 830, 164, 1985.
7. **Wayne, W. I. and Fruton, J. S.,** Thermolysin-catalyzed peptide bond synthesis, *Proc. Natl. Acad. Sci. U.S.A.,* 80, 3241, 1983.
8. **Oyama, K., Kihara, K.-I., and Nonaka, Y.,** On the mechanism of the action of thermolysin: kinetic study of the thermolysin-catalyzed condensation reaction of N-Benzyloxycarbonyl-L-aspartic acid with L-phenylalanine methyl ester, *J. Chem. Soc. Perkin Trans.,* 2, 356, 1981.
9. **Bozler, H., Wayne, S. I., and Fruton, J. S.,** Specificity of pepsin-catalyzed peptide bond synthesis, *Int. J. Peptide Protein Res.,* 20, 102, 1982.
10. **Fruton, J. S.,** The specificity and mechanism of pepsin action, *Adv. Enzymol. Relat. Areas Mol. Biol.,* 33, 401, 1970.
11. **Ardelt, W. and Laskowski, M., Jr.,** Thermodynamics and kinetics of the hydrolysis and resynthesis of the reactive site peptide bond in turkey ovomucoid third domain by aspergillopeptidase B, *Acta Biochim. Polonica,* 30, 115, 1983.
12. **Quast, U., Engel, J., Steffen, E., Tschesche, H., and Kupfer, S.,** Kinetics of the interaction of α-chymotrypsin with trypsin kallikrein inhibitor (Kunitz) in which the reactive-site peptide bond Lys-15—Ala-16 is split, *Eur. J. Biochem.,* 86, 353, 1978.
13. **Estell, D. A., Wilson, K. A., and Laskowski, M., Jr.,** Thermodynamics and kinetics of the hydrolysis of the reactive-site peptide bond in pancreatic trypsin inhibitor (Kunitz) by Dermasterias imbricata trypsin 1, *Biochemistry,* 19, 131, 1980.
14. **Mattis, J. A. and Laskowski, M., Jr.,** pH dependence of the equilibrium constant for the hydrolysis of the Arg$_{63}$-Ile reactive-site peptide bond in soybean trypsin inhibitor (Kunitz), *Biochemistry,* 12, 2239, 1973.
15. **Markussen, J. and Vølund, A.,** Kinetics of tryptic transpeptidation of insulins, in *Proc. 8th Am. Peptide Symp.,* Hruby, V. and Rich, D. H., Eds., Pierce Chem., Rockford, Ill., 1984, 207.
16. **Markussen, J. and Vølund, A.,** Kinetics of trypsin catalysis in the industrial conversion of porcine insulin to human insulin, in *Enzymes in Organic Synthesis,* Symp. 111, Porter, R. and Clark, S., Eds., Ciba Foundation, Pitman, London, 1985, 188.
17. **Lineweaver, H. and Burk, D.,** The determination of enzyme dissociation constants, *J. Am. Chem. Soc.,* 56, 658, 1934.
18. **Kullmann, W.,** Enzymatic synthesis of Leu- and Met-enkaphalin, *Biochem. Biophys. Res. Commun.,* 91, 693, 1979.
19. **Kullmann, W.,** Proteases as catalysts of enzymic syntheses of opioid peptides, *J. Biol. Chem.,* 255, 8234, 1980.
20. **Kullmann, W.,** Kinetics of chymotrypsin- and papain-catalyzed synthesis of (leucine)enkephalin and (methionine)enkephalin, *Biochem. J.,* 220, 405, 1984.
21. **Walsh, C.,** *Enzymatic Reaction Mechanisms,* Freeman, Cooper and Co., San Francisco, 1979, 53.
22. **Frère, J. M.,** Enzymic mechanims involving concomitant transfer and hydrolysis reactions, *Biochem. J.,* 135, 469, 1973.
23. **Folk, J. E.,** Mechanism of action of guinea pig liver transglutaminase, *J. Biol. Chem.,* 244, 3707, 1969.
24. **Cleland, W. W.,** The kinetics of enzyme-catalyzed reactions with two or more substrates or products. II. Inhibition: nomenclature and theory, *Biochim. Biophys. Acta,* 67, 173, 1963.

25. **Fromm, H. J.,** The use of alternative substrates in studying enzymic mechanisms involving two substrates, *Biochim. Biophys. Acta,* 81, 413, 1964.
26. **Fromm, H. J.,** Initial rate enzyme kinetics, in *Molecular Biology, Biochemistry, and Biophysics,* Vol. 22, Kleinzeller, A., Springer, G. F., and Wittmann, H. G., Eds., Springer-Verlag, Berlin, 1975, 152.
27. **Segel, I. W.,** *Enzyme Kinetics,* John Wiley and Sons, New York, 1975, 793.
28. **Schechter, I. and Berger, A.,** On the active site of proteases. III. Mapping the active site of papain; specific peptide inhibitors of papain, *Biochem. Biophys. Res. Commun.,* 32, 898, 1968.

Chapter 12

PROTEASES AS CATALYSTS IN PROTECTING GROUP CHEMISTRY

I. INTRODUCTION

The use of proteases in synthetic peptide chemistry is not confined solely to the basic step of peptide bond formation. In addition, proteases have also been described as useful and efficient agents for the introduction and removal of protecting groups. This area of synthetic methodology is by no means marginal in interest; indeed as noted previously (cf. Chapter 6, Section I.B and Chapter 8, Section II), a judicious protecting group strategy is indispensable for a successful peptide synthesis. Consequently, a major part of peptide synthetic chemistry has been concerned with the development of suitable protecting groups,[1,2] which are commonly subdivided into two categories: semipermanent and temporary blocking groups. To recap briefly, the so-called semipermanent protector groups have to survive the multiple reaction steps occurring during the progress of synthesis and their final removal usually represents the last step of the synthetic pathway, i.e., the release of the free target peptides. In contrast, temporary protector groups must be removed prior to each elongation step of the growing peptide chain. Criteria for appropriate semipermanent blocking groups are readily categorized as follows: convenient introduction, inertness to the conditions prevailing during synthetic manipulations, and reversibility without affecting the integrity of the final product. Satisfactory temporary protecting groups must be selectively removable under mild conditions without injuring semipermanent blocking groups or, as a matter of course, the peptidic backbone. The traditional procedure of unidirectional, stepwise incorporation of single amino acid derivatives involves the removal of the N^α-protecting group prior to each elongation cycle. In cases where the synthetic strategy involves fragment coupling as well, the α-carboxyl protection has also to be a selectively removable in the presence of semipermanent and temporary N^α-protector groups.

In principle, the demand of selectivity and the mildest possible conditions of both introduction and removal of blocking groups can be best met by enzymatic procedures. While the enzymatic manipulation of protecting groups is still in its infancy some remarkable studies on this subject have already been published. These reports can be divided into the enzymatic introduction and removal of either main-chain-, or N^α-, or α-carboxyl protecting groups, or side-chain protecting groups.

II. N^α-PROTECTING GROUPS

The first application of enzymes for the removal of α-amino blocking group was reported by Holley.[3] The blocking agents most commonly used in chemical syntheses, such as urethane derivatives, are usually not effective substrates for the proteases. The author circumvented this problem by chosing a protease-labile blocking entity, namely Bz-Phe-OH, which represent a composite of the well-known N^α-blocking groups and a chymotrypsin-sensitive amino acid residue. The penultimate step of the synthetic route, finally leading to the desired dipeptide H-Leu-Leu-OH, yielded Bz-Phe-Leu-Leu-OH. The last step, the removal of Bz-Phe-OH — the "N^α-protecting group" — was catalyzed by α-chymotrypsin and gave the dipeptide leucylleucine in 80% yield. Obviously, this approach is restricted to peptides lacking other chymotrypsin-labile bonds. An analogous methodology was adopted by Meyers and Glass, who used the trypsin-sensitive Z-Arg-moiety as an N^α-protecting group during the synthesis of oxytocin fragments.[4] The respective amino acid residue (X) to be added was incorporated as an integral part of the corresponding ortho-nitrophenyl Z-Arg-aminoacylate (Z-Arg-X-ONp). The coupling of the activated dipeptide unit to the growing peptide chain was followed subsequently by the removal of the Z-Arg-protection via tryptic hydrolysis at pH 8. A similar approach to the stepwise synthesis of deamino-oxytocin by addition of Z-Arg-X-ONp units to the growing peptide chain and followed by tryptic removal of Z-Arg-OH was presented by Glass.[5] Unfortunately, the introduction of dipeptide units into a peptide chain, which is already a circuitous procedure, represents a fragment condensation step which is generally prone to racemization. Consequently, it is questionable, whether the advantage taken from the mildest possible removal conditions can compensate for the enhanced risk of destroying the chiral integrity of the expected products. However, this problem can be eliminated by using a protease to stereospecifically catalyze the peptide-bond-forming step. The validity of this approach was demonstrated by Widmer et al.[6] who synthesized an undecapeptide fragment of the epidermal growth factor (cf. Chapter 8, Section IV, Figure 13). The authors used Bz-Phe- and Bz-Arg moieties as protease-labile, temporary protector groups which could eventually be removed by chymotryptic and tryptic catalysis, respectively. The trypsin-sensitive Bz-Arg unit was used as a semipermanent N^α-protecting group during the enzymatic synthesis of Met-enkephalin (cf. Chapter 8, Section II, Figure 2). The exopeptidase CPD-Y was employed to catalyze the peptide-bond-forming steps, thereby permitting the peptide chain elongation from the N- to the C-terminus without provoking either product racemization or premature removal of the semipermanent N^α-protection. Bz-Arg-OH was finally removed by tryptic hydrolysis to yield the target peptide. Of course, this procedure was only feasible because the enkephalins are devoid of any trypsin-labile bond in their sequence.

The exploitation of the hydrolytic potential of enzymes other than proteases for the enzymatic removal of N^α-blocking groups merits particular attention. In this connection, Widmer et al.[8] suggested the use of penicillin acylase, which is known to cleave phenylacetyl moieties from α-amino groups of amino acids[9,10] and from the ϵ-amino group of lysine,[11] as catalyst for N^α-deblocking steps. To permit the more general application of this technique, it is worth looking for further acylases or related hydrolases, and also for alternative protecting groups. Given that N^α-acylated amino acids run the risk of being racemized during chemical coupling steps, enzymes should be used to catalyze not only the introduction and removal of the blocking agents but also, where possible, the peptide-bond-forming step. This was certainly feasible during the synthesis of the previously mentioned undecapeptide fragment of the mouse epidermal growth factor.[6] Here Bz-Arg- and Bz-Phe moieties were introduced as temporary N^α-protections, respectively, by tryptic and chymotryptic catalysis. With one exception, the subsequent coupling steps were catalyzed by proteases. The final release of the protease-labile protecting groups from the newly formed peptides were accomplished again in the presence of trypsin or chymotrypsin, respectively (cf. Chapter 8, Section II, Figure 13). Similarly the Bz-Arg-protecting group was tryptically introduced and removed during the synthesis of Met-enkephalin (cf. Chapter 8, Section II).[7]

III. α-CARBOXYL PROTECTION

The use of protease-sensitive, α-carboxyl-protecting groups represents a more versatile strategy than that of protease-labile, α-amino-blocking groups, primarily because the esterase activity possessed by most the proteases readily enables the mild and convenient removal of a variety of α-carboxyl esters. Cleavage of an alkyl ester bond is an often used reaction during synthetic peptide chemistry. However, this procedure frequently gives rise to undesired side reactions,[1] in particular, where peptides of advanced chain length require prolonged reaction times.

It has been known since 1948 that in particular α-chymotrypsin and trypsin, in addition to their proteolytic potential, are capable of cleaving ester linkages.[12,13] According to Schwert and Eisenberg[14] the chemical nature of the alkyl esters did not affect their ability to be cleaved by trypsin. Indeed methyl-, ethyl-, and *t*-butyl esters were readily hydrolyzed. In 1962 Walton et al.[15] reported the successful α-chymotrypsin-mediated removal of a methyl ester from the C-terminal phenylalanine residue of an isoleucine-5-angiotensin octapeptide. The authors observed that the ester bond was selectively and completely split within 2 hr at pH 6.0. In contrast to this, acid- or base-catalyzed demethylation resulted in rather low yields. The potentially α-chymotrypsin-labile, tyrosyl-isoleucine bond of angiotensin was not subject to peptidase activity at an enzyme to substrate ratio of 1:10.[4] Additionally, the methyl esters of the following peptides could be efficiently removed by α-chymotrypsin catalysis: Z-Pro-Phe-OMe (90%); Z-Val-Tyr-OMe (80%); Z-Arg-(NO$_2$)-Val-Tyr-OMe (95%); and Z(CH$_2$)-Asp-Tyr-OMe (80%).

The effect of α-chymotrypsin and trypsin on the hydrolysis of a variety of peptide esters was systematically explored by Kloss and Schröder.[16] As might be expected, the C-terminal esters of phenylalanine, tyrosine, and tryptophan were easily hydrolyzed by chymotrypsin, but surprisingly the corresponding esters of threonine, serine, and histidine were rapidly cleaved as well. In contrast, alanine, valine, and leucine esters were cleaved less efficiently. In the cases of aspartic- and glutamic-acid diesters, only the α-carboxyl esters were removed. In addition to its effect on arginine and lysine esters, trypsin also readily cleaved phenyl-alanine-, tyrosine-, histidine-, threonine-, serine-, and ε-formulated lysine esters. The be-havior of trypsin with respect to esters of alanine, valine, and leucine, resembled that of α-chymotrypsin, the esters being cleaved to a lesser extent. On the other hand, D-amino acid esters were rather poor substrates for tryptic hydrolysis. Apparently then, the esterase spec-ificity of α-chymotrypsin and trypsin covers a broader spectrum than their peptidase spec-ificities. Furthermore, this "versatility" of the two proteases provides an opportunity to use either trypsin or α-chymotrypsin for the removal of a given α-carboxyl protecting group. This alternative may be of particular use where the peptide to be synthesized contains a peptide bond, sensitive to one or another, although not both, of these proteases.

The applicability of an alkaline protease of B. subtilis DY to the hydrolysis of several amino acid and peptide esters was described by Aleksiev et al.[17] The protease mediated the cleavage of methyl-, ethyl-, and benzyl esters at pH 8 usually giving yields of greater than 90%. *t*-Butyl esters, however, were not hydrolyzed. In general, the alkaline protease ex-hibited a broad structural specificity, however, if did not cleave the γ-ethyl ester of glutamic acid. Furthermore, the integrity of potentially susceptible peptide bonds was not impaired because under the experimental conditions used, the rate of ester hydrolysis was five orders of magnitude larger than that of proteolysis. The efficiency of the alkaline protease in the cleavage of ester bonds exceeded that achieved by both α-chymotrypsin and by alkaline saponification.

In another study, Hermann and Salewski[18] reported the enzymatic cleavage of C-terminal ester bonds via thermitase-mediated catalysis. In the presence of β- and γ-esters of amino dicarboxylic acids, this alkaline serine protease, derived from *Thermoactinomyces vulgaris*, removes selectively α-alkyl- and benzyl ester groups from peptide derivatives, provided that the C-terminal amino acid is of the L-configuration. Due to its broad substrate specificity, thermitase is more widely applicable to problems of protecting group removal than either trypsin or chymotrypsin. At a pH optimum of 8.0, the esterase activity can be preferentially enhanced over peptidase activity by either decreasing the concentration ratio of enzyme to substrate or by increasing the percentage of organic co-solvents.

The esterase potential of CPD-Y at basic pH values (cf. Chapter 7, Section VII) was taken advantage of by Royer and Anantharamaiah[19] and by Royer et al.[20] for the sequential synthesis in aqueous solution of a number of model peptides. Following the procedure described by Mutter et al.,[21] the peptide chain was chemically synthesized on carboxymethyl polyethyleneglycol starting from the N-terminal end. Prior to each elongation step the removal of the temporary ethyl ester protection of the α-carboxyl groups was achieved using immobilized sepharose-bound CPD-Y at pH 8.5. This approach takes advantage of the stereoselectivity of the protease, since none of the peptide esters terminating with D-amino acids is recognized as substrate; i.e., they remain blocked and do not participate in further coupling steps. In addition to its stereoselectivity, CPD-Y also acts regiospecifically, so that β-esters of aspartic acid derivatives are not hydrolyzed. In addition to the esterase potential of CPD-Y its amidase activity has also been exploited for the removal of α-carboxyl blocking groups. During the CPD-Y catalyzed synthesis of Met-enkephalin,[7] the C-terminal end of the growing peptide chain was temporarily protected by an amide group, which was subsequently removed by CPD-Y-catalysis prior to chain elongation (cf. Chapter 8, Section II, Figure 2). This and a similar procedure used during the synthesis of a fragment of the mouse epidermal growth factor[6] (cf. chapter 8, section IV, Figure 13) represent two examples of the rare use of an amide group to serve as an α-carboxyl protecting group. This method is normally avoided since this blockage cannot be removed by chemical means without detrimental side reactions, although of course protease-mediated removal circumvents this pitfall.

An arginine residue bound to a polymeric carrier (polyethyleneimine) was used by Blecher and Pfaender[22] as a permanent α-carboxyl "protection" during the synthesis of the model peptide H-Ala-Trp-Ile-OH. The peptide was first assembled on the arginyl-resin by a stepwise manner and then treated with trypsin to cleave the tetrapeptide H-Ala-Trp-Ile-Arg-OH from the carrier. Following this, the C-terminal arginine residue was removed through the hydrolytic action of the arginine-specific carboxylpeptidase B in order to release the free target tripeptide. In an analogous fashion, a thermolysin-sensitive α-carboxyl protection, given by the tripeptide H-Leu-Gly-Gly-OEt, was used by Ohno and Anfinsen to prepare a partially protected analogue of the staphyloccal nuclease sequence 43 to 50.[23] This strategy was selected, because the removal of a C-terminal ester group via alkaline saponification destabilized some side-chain blocking groups and possibly the chiral integrity of the peptide. The bond, which was hydrolyzed by thermolysin at pH 7.4, was a glycyl-leucine linkage.

Enzymic rather than chemical means have also been used for the detachment of other protected peptides from polymeric supports. The proteases employed in these cases were thermolysin and α-chymotrypsin.[24] Considering the fact that side reactions are often encountered during the chemical removal of peptides from their solid supports, the enzymic approach, when applicable, offers a valuable alternative to chemical methods.

The deprotection step is not the only source of side reactions in synthetic peptide chemistry. Undesired by-products may also be created during the introduction of protecting groups. This problem can be overcome most efficiently, if the introduction of protecting groups is also carried out via enzyme-catalysis. While to the best of the author's knowledge, the abovementioned synthesis of the epidermal growth factor[6] and of Met-enkephalin[7] are the only examples of enzymatically engineered Nα-protection, the masking of α-carboxyl groups by means of proteases has been reported more frequently. For example, the esterification of the α-carboxyl group of Nα-Ac-Tyr-OH was described by Ingalls et al.[25] Initially using free α-chymotrypsin as catalyst in a 1:1 by volume aqueous-organic solvent system, the authors found that the yield of ester formation was below 30% However, while reduction of the water concentration by supplying additional organic co-solvents (ethanol-glycerol) (1:1 by volume), led to inactivation of the free enzyme, in the presence of 80% nonaqueous solvents the efficiency of esterification by immobilized α-chymotrypsin could be increased to more than 50% at pH 7.6. Analysis of the reaction mixture demonstrated that the major product was the cognate ethyl ester but glyceryl ester had also been formed in significant amounts. On the other hand, freely mobile subtilisin Carlsberg (subtilopeptidase A) was an effective catalyst for this reaction even if the water concentration was reduced to 2%. Maximum yields of esterification (33 to 38%) were achieved at a pH of 4.6 in the presence of only 5: to 10% water. In contrast to the behavior of α-chymotrypsin, immobilized subtilisin was unable to catalyze the formation of the desired esters of Nα-Ac-Tyr-OH.

Further investigations of α-chymotrypsin-catalyzed esterification techniques were reported by Martinek et al.,[26] Tarquis et al.,[27] and Vidaluc et al.[28] Ethylation of N^α-Bz-Phe-OH in the presence of ethanol, was accomplished by using a suspension of silica gel impregnated with an aqueous solution of α-chymotrypsin in chloroform at pH 7. The product yield was 100% in a biphasic system composed of chloroform-water (95:5) (v/v). The formation of N^α-Ac-Tyr-OEt from N^α-Ac-Tyr-OH and ethanol was achieved in a biphasic aqueous-organic system using α-chymotrypsin covalently coupled to a porous aminated silica gel resin with a specific area of 24 m^2/g (particle diameter between 100 and 200 μm).[27] The reaction was initially conducted at pH 6.8 in a mixture of 0.1 M citrate-phosphate buffer and chloroform (5.6:94.4) (v/v), and N^α-Ac-Tyr-OEt was obtained in a 35% yield. However, optimization of the esterification procedure resulted in a remarkably higher yield (85 to 90%). In fact the authors observed that maximum yields could be obtained at pH 4.2 by increasing the specific area to 50 m^2/g and by using 1 M citrate-phosphate buffer as aqueous phase.

Besides chymotrypsin and subtilisin, the protease papain has also been used to esterify N^α-protected amino acids. Thus, the methyl or ethyl esters of some 20 Boc- or Z-protected amino acids have been prepared in the presence of papain by Pinheiro-Da-Silva et al.[29]

In earlier studies on protease-catalyzed peptide bond formation, the α-COOH-protection of amine components had been accomplished by the anilide group.[30,31] Anilides were readily obtained by incubating the respective N^α-protected amino acid with aniline in the presence of papain.[30] In a comparative investigation, Fox et al.[32] show that the efficiency of the papain-mediated introduction of the anilide group was in the following order (pH, ca. 5.0): N^α-benzoylated leucine, -alanine, -glycine, and -valine. In another report, Carty and Kirschenbaum showed[33] that chymopapain was 20% more active than papain in promoting the formation of Bz-Gly-NHPh from Bz-Gly-OH and aniline. The anilide protection of α-carboxyl functions may actually have favored the synthesis of model peptides by reducing their solubility. However, in general this kind of protection is completely inadequate because it lacks one essential requirement of a suitable temporary or permanent protecting group, namely the selective reversibility of reaction. Indeed anilides cannot be chemically removed without seriously impairing the integrity of the newly formed peptide chain.

An alternative α-COOH protecting group, the phenylhydrazide moiety, which is chemically related to the anilides, can also be introduced by papain catalysis.[34] In contrast to the anilides however, the phenylhydrazides are readily and selectively removable by oxidizing agents such as ferric chloride or copper acetate.[35] The preparation by papain-catalysis of a series of different N^α-acyl amino acid derivatives using phenylhydrazine C-terminal protection has been reported by Milne and Stevens[36] and Milne and Carpenter.[37] Beyond that, this method has also been successfully applied to the synthesis of the enkephalins and cholecystokinin-related peptides.[38,39] With the exception of proline, the α-carboxylate phenylhydrazides of all codogenous N^α-protected amino acids have been selectively synthesized via papain catalysis by Čeřovský and Jošt.[40] Although there were significant differences in reaction yields between the respective amino acid derivatives, this study demonstrated the facility of papain as an almost universal condensation reagent. In this work amino acids bearing benzyloxycarbonyl- and t-butyloxycarbonyl N^α-protector groups, respectively, were used; however, N^α-9-fluorenylmethyloxycarbonylamino acids did not react with phenylhydrazide in the presence of papain. In addition to papain, trypsin has also been used to catalyze the conversion of peptides to the corresponding peptide phenylhydrazides.[41] Furthermore, trypsin will also catalyze the introduction of esters, hydrazides, and substituted hydrazides, such as Boc-NHNH$_2$ and Z-NHNH$_2$ into α-COOH groups.[41,42]

In addition to their primary function as effective carboxyl-protecting groups, phenylhydrazides can be easily oxidized in nonaqueous media by N-bromosuccinimide to provide the respective phenyldiimides, which represent strong acylating agents.[43] These hydrazo compounds readily react with nucleophiles such as alcohols, amines, or water. During the syntheses of the enkephalins[38] and of a cholecystokinin octapeptide amide,[39] several peptide phenylhydrazides were rearranged to yield the corresponding ethyl esters. These in turn served as acyl donors for α-chymotrypsin-catalyzed couplings or to a peptide amide, which represented an intermediate of the cholecystokinin octapeptide amide (cf. Chapter 8, Section III, Figure 5). A similar procedure was also applied by Canova-Davis and Carpenter to the semisynthesis of bovine insulin.[44] The authors coupled via trypsin-catalysis the carboxyl group of arginine B_{22} in N^{A1}, $-N^{B1}$-(Boc_2)-DOI with phenylhydrazine. The resulting phenylhydrazide derivative was oxidized to yield the corresponding phenyldiimide, which was then subsequently reacted with the (ϵ-Boc)-(B_{23-30}) fragment of bovine insulin to give $N^{\alpha A1}$, $-N^{\alpha B1}$, $N^{\epsilon B29}$, $-(Boc_3)$ insulin. In this case however, the phenylhydrazide was employed as an intermediate in the semisynthesis of a peptide rather than a blocking group.

IV. ENZYME-SENSITIVE SIDE-CHAIN PROTECTION

Studies dealing with the enzymatic removal of side-chain protecting groups have been reported in relatively few instances. Indeed, so far as side-chain carboxyl groups are concerned, there seems to exist only a single report. The authors of this study attempted to ameliorate the lack of suitable blocking groups for the side-chain functionalities of amino dicarboxylic acids during semisyntheses by the following strategy:[45] the β- or γ-carboxylate, respectively, of aspartic- or glutamic acid, is temporarily blocked by an arginine methyl ester which, when the occasion demands, can be removed by successive treatment with trypsin and CPD-B (carboxypeptidase B). The practicability of this concept was explored by protecting the β- and γ-carboxyl functions of the model compounds N^{α}-Z-isoasparagine and -isoglutamine, which can serve as substitutes for internal aspartic or glutamic acid residues, with arginine methyl esters. The "side-chain" protection of the isomeric forms of Z-Asn-OH and Z-Gln-OH could be removed by tryptic hydrolysis of the methyl ester followed by the release of free arginine via CPD-B-catalysis. The trypsin-controlled ester cleavage proceeded very rapidly, whereas the CPD-B-mediated hydrolysis of the β- and in particular, of the γ-peptide bond, was slow compared to that expected for the corresponding α-peptide linkages. Admittedly, this approach may interfere with additional trypsin-labile bonds within the peptide being synthesized, but its applicability may be broadened by the use of proteases such as clostripain, the specificity of which is strictly confined to arginine residues.[46]

Furthermore, some reports have been published which describe the enzymatic deblocking of the lysine side-chain function. Jering et al.[47] used the porcine acyl-lysine deacylase to remove an acetyl group from the ϵ-amino function of a lysine residue within a chemically modified trypsin kallikrein inhibitor. In a different study, which dealt with the synthesis of deamino-lysine vasopressin, Brtnik et al.[11] reported the use of the phenylacetyl moiety as N^{α}-protection of the Lys_8-side chain. This protector group was selectively removable via penicillin-acylase catalysis, without injuring the peptidic backbone.

In addition to the aforementioned enzyme-sensitive protections for the third functionality of lysine residues, Glass and Pande[48] introduced the pyroglutamyl unit as enzymatically removable N^{α}-protection. Selective cleavage of this blocking group, i.e., the release of pyroglutamic acid, was accomplished by the catalytic action of pyrrolidonecarboxylpeptidase from calf liver. The practicability of this procedure was demonstrated by the successful removal via pyrrolidonecarboxylpeptidase catalysis of the pyroglutamyl residue from the side chain of certain lysine units which were integral parts not only of model compounds but also of larger peptides such as RNase A and its S-peptide.

REFERENCES

1. **Wünsch, E., Ed.,** *Houben-Weyl, Methoden der organischen Chemie,* Vol. 15, Part I, Thieme, Stuttgart, 1974.
2. **Gross, E. and Meienhofer, J., Eds.,** *The Peptides: Analysis, Synthesis, Biology,* Vol. 3, Academic Press, New York, 1981.
3. **Holley, R. W.,** Enzymatic removal of the protecting group in peptide synthesis, *J. Am. Chem. Soc.,* 77, 2552, 1955.
4. **Meyers, C. and Glass, J. D.,** Enzymes as reagents in peptide synthesis: enzymatic removal of amine protecting groups, *Proc. Natl. Acad. Sci. U.S.A.,* 72, 2193, 1975.
5. **Glass, J. D.,** Enzymes as reagents in the synthesis of peptides, *Enzyme Microb. Technol.,* 3, 2, 1981.
6. **Widmer, F., Bayne, S., Houen, G., Moss, B. A., Rigby, R. D., Whittacker, R. G., and Johansen, J. T.,** Use of proteolytic enzymes for synthesis of fragments of mouse epidermal growth factor, in *Peptides 1984, Proc. 18th Eur. Peptide Symp.,* Regnarsson, U., Ed., Almquist and Wiksell International, Stockholm, 1984, 193.
7. **Widmer, F., Breddam, K., and Johansen, J. T.,** Carboxypeptidase Y as a catalyst for peptide synthesis in aqueous phase with minimal protection, in *Peptides 1980, Proc. 16th Eur. Peptide Symp.,* Brunfeldt, K., Ed., Scriptor, Copenhagen, 1981, 46.
8. **Widmer, F., Ohno, M., Smith, M., Nelson, N., and Anfinsen, C. B.,** Enzymatic peptide synthesis, in *Peptides 1982, Proc. 17th Eur. Peptide Symp.,* Blahá, K. and Maloň, P., Eds., Walter de Gruyter, Berlin, 1983, 375.
9. **Cole, M.,** Properties of the penicillin deacylase enzyme of *Escherichia coli, Nature (London),* 203, 519, 1964.
10. **Kaufmann, W. and Bauer, K.,** Variety of substrates for a bacterial benzyl penicillin-splitting enzyme, *Nature (London),* 203, 520, 1964.
11. **Brtnik, F., Barth, T., and Jošt, K.,** Use of phenacetyl group for protection of the lysine N^ε-amino group in synthesis of peptides, *Collect. Czech. Chem. Commun.* 46, 1983, 1981.
12. **Kaufman, S., Schwert, G. W., and Neurath, H.,** The specific peptidase and esterase activities of chymotrypsin, *Arch. Biochem.,* 17, 203, 1948.
13. **Schwert, G. W., Neurath, H., Kaufman, S., and Snoke, J. E.,** The specific esterase activity of trypsin, *J. Biol. Chem.,* 172, 221, 1948.
14. **Schwert, G. W. and Eisenberg, M. A.,** The kinetics of the amidase and esterase activities of trypsin, *J. Biol. Chem.,* 179, 665, 1949.
15. **Walton, E., Rodin, J. O., Stammer, C. H., and Holly, F. W.,** Peptide synthesis. An application of the esterase activity of chymotrypsin, *J. Org. Chem.,* 27, 2255, 1962.
16. **Kloss, G. and Schröder, E.,** Über enzymatische Verseifung von Peptid-Estern, *Hoppe-Seyler's Z. Physiol. Chem.,* 336, 248, 1964.
17. **Aleksiev, B., Schamilian, P., Widenow, G., Stoev, S., Zachariev, S., Golovinsky, E.,** Verwendung von alkalischer Protease des Bacillus-subtilis-Stammes DY zur Hydrolyse von Aminosäuve und Peptid-Estern, *Hoppe-Seyler's Z. Physiol.Chem.,* 362, 1323, 1981.
18. **Hermann, P. and Salewski, L.,** Thermitase-catalyzed hydrolysis of the C-terminal ester group of protected peptide ester derivatives, in *Peptides 1982, Proc. 17th Eur. Peptide Symp.,* Blahá, K. and Maloň, P., Eds., Walter de Gruyter, Berlin, 1983, 39.
19. **Royer, G. P. and Anantharamaiah, G. M.,** Peptide synthesis in water and the use of immobilized carboxipeptidase Y for deprotection, *J. Am. Chem. Soc.,* 101, 3394, 1979.
20. **Royer, G. P., Hsiao, H. Y., and Anantharamaiah, G. M.,** Use of immobilized carboxypeptidase Y (I-CPY) as a catalyst for deblocking in peptide synthesis, *Biochemie,* 62, 537, 1980.
21. **Mutter, M., Hagenmaier, H., and Bayer, E.,** Eine neue Methode zur Synthese von Polypeptiden, *Angew. Chem.,* 83, 883, 1971.
22. **Blecher, H. and Pfaender, P.,** Peptidsynthesen in wässriger Phase. I. Polyäthylenimin als wasserlöslicher Träger bei der *N*-carboxyanhydrid-Methode, *Justus Liebigs Ann. Chem.,* 1263, 1973.
23. **Ohno, M. and Anfinsen, C. B.,** Partial enzymic deprotection in the synthesis of a protected octapeptide bearing a free terminal carboxyl group, *J. Am. Chem. Soc.,* 92, 4098, 1970.
24. **Anfinsen, C. B.,** Characterization of Staphylococcal nuclease and the status of studies on its chemical synthesis, *Pure Appl. Chem.,* 17, 461, 1968.
25. **Ingalls, R. G., Squires, R. G., and Butler, L. G.,** Reversal of enzymatic hydrolysis: rate and extent of ester synthesis as catalyzed by chymotrypsin and subtilisin Carlsberg at low water concentrations, *Biotechnol. Bioeng.,* 17, 1627, 1975.
26. **Martinek, K., Semenov, A. N., and Berezin, I. V.,** Enzymatic synthesis in biphasic aqueous organic systems. I. Chemical equilibrium shift, *Biochim. Biophys. Acta,* 658, 76, 1981.
27. **Tarquis, D., Monsan, P., and Durand, G.,** Synthèse enzymatique d'esters d'acides aminés en milieu organique, *Bull. Soc. Chim. Fr.,* 2, 76, 1980.

28. **Vidaluc, J. L., Baboulene, M., Speziale, V., Lattes, A., and Monsan, P.,** Optimization of the enzymatic synthesis of amino acid esters. Reaction in polyphasic medium, *Tetrahedron,* 39, 269, 1983.

29. **Pinheiro-Da-Silva F., L., Dovicchi, J. C. L., Muradian, J., Seidel, W. F., and Tominaga, M.,** Papain-catalyzed esterification of *N*-tert-butyl-oxycarbonyl and *N*-benzyloxycarbonyl amino acids, *Brazilian J. Med. Biol. Res.,* 14, 373, 1981.

30. **Bergmann, M. and Fraenkel-Conrat, H.,** The enzymatic synthesis of peptide bonds, *J. Biol. Chem.,* 124, 1, 1938.

31. **Bergmann, M. and Fruton, J. S.,** Some synthetic and hydrolytic experiments with chymotrypsin, *J. Biol. Chem.,* 124, 321, 1938.

32. **Fox, S. W., Pettinga, C. W., Halverson, J. S., and Wax, H.,** Enzymatic synthesis of peptide bonds. II. "Preferences" of papain within the monoaminomonocarboxylic acid series, *Arch. Biochem.,* 25, 21, 1950.

33. **Carty, R. P. and Kirschenbaum, D. M.,** The papain-catalyzed synthesis of hippuryl anilide I. General properties of the enzyme system, *Biochim. Biophys. Acta,* 85, 446, 1964.

34. **Bergmann, M. and Fraenkel-Conrat, H.,** The role of specificity in the enzymatic synthesis of proteins. Syntheses with intracellular enzymes, *J. Biol. Chem.,* 119, 707, 1937.

35. **Milne, H. B., Halver, J. E., Ho, D. S., and Mason, M. S.,** The oxidative cleavage of phenylhydrazide groups from carboallyloxy-α-amino acid phenylhydrazides and carboallyloxydipeptide phenylhydrazides, *J. Am. Chem. Soc.,* 79, 637, 1957.

36. **Milne, B. H. and Stevens, C. M.,** Amino acid derivatives. II. Enzymatic synthesis of phenylhydrazides of carboallyloxyamino acids, *J. Am. Chem. Soc.,* 72, 1742, 1950.

37. **Milne, B. H. and Carpenter, F. H.,** Peptide synthesis via oxidation of *N*-acyl-α-amino acid phenylhydrazides. III. Dialanyl-insulin and diphenylalanyl-insulin, *J. Org. Chem.,* 33, 4476, 1968.

38. **Kullmann, W.,** Proteases as catalysts for enzymic syntheses of opioid peptides, *J. Biol. Chem.,* 255, 8234, 1980.

39. **Kullmann, W.,** Protease-catalyzed peptide bond formation: application to synthesis of the COOH-terminal octapeptide amide of cholecystokinin, *Proc. Natl. Acad. Sci. U.S.A.,* 79, 2840, 1982.

40. **Čeřovský, V. and Jošt, K.,** Papain-catalyzed synthesis of phenylhydrazides of *N*-acyl amino acids, *Collect. Czech. Chem. Commun.,* 49, 2557, 1984.

41. **Jones, R. M. L. and Offord, R. E.,** The proteinase-catalyzed synthesis of peptide hydrazides, *Biochem. J.,* 203, 125, 1982.

42. **Yagisawa, S.,** Studies on protein semisynthesis. I. Formation of esters, hydrazides, and substituted hydrazides of peptides by the reverse reaction of trypsin, *J. Biochem.,* 89, 491, 1981.

43. **Milne, H. B. and Kilday, W.,** Peptide synthesis via oxidation of *N*-acetyl-α-amino acid phenylhydrazides, *J. Org. Chem.,* 30, 64, 1965.

44. **Canova-Davis, F. and Carpenter, F. H.,** Semisynthesis of insulin: specific activation of the arginine carboxyl group of the B chain of desoctapeptide-(B23-30)-insulin (bovine), *Biochemistry,* 20, 7053, 1981.

45. **Glass, J. and Pelzig, M.,** Enzymes as reagents in peptide synthesis: enzyme-labile protection for carboxyl groups, *Proc. Natl. Acad. Sci. U.S.A.,* 74, 2739, 1977.

46. **Mitchel, W. M. and Harrington, W. F.,** Purification and properties of a clostridiopeptidase B (clostripain), *J. Biol. Chem.,* 23, 4683, 1980.

47. **Jering, H., Schorp, G., and Tschesche, H.,** Enzymatic cleavage of the N$^{\alpha}$-acetyl protecting group from lysine in proteins (Kunitz-trypsin-kallikrein-inhibitor) in vitro and in vivo, *Hoppe-Seyler's Z. Physiol. Chem.,* 355, 1129, 1974.

48. **Glass, J. D. and Pande, C. S.,** Pyroglutamyl groups as enzyme-labile protection of the lysine side-chain, in *Peptides — Structure and Function, Proc. 8th Am. Peptide Symp.,* Hruby, V. and Rich, D. H., Eds., Pierce Chemical, Rockford, Ill., 1984, 203.

Chapter 13

SHORTCOMINGS AND ALTERNATIVES

Despite its promising features, the enzymatic approach to peptide synthesis has yet to reach general applicability. A major limitation of the method is determined by the lack of a complete set of proteases whose stringent specificity would enable the synthetic chemist to prepare any conceivable peptide linkage without interfering with pre-existing bonds. Bearing in mind that the 20 genetically encoded amino acid residues can participate in 400 different types of peptide bonds, it seems unlikely that a complete "arsenal" of proteases with exclusive specificity for all possible peptide bonds can ever be assembled. Indeed the broad specificity demonstrated by many of the proteases normally involved in protein degradation in vitro, makes it unlikely that each bond is selectively cleaved by a unique protease. Even if one takes in addition those proteases serving various regulatory functions, nature still fails to provide the peptide chemist with the desired collection of highly selective proteases.

During their initial studies on enzymatic peptide synthesis, Bergmann and Fraenkel-Conrat encountered problems resulting from the relative lack of specificity of a protease.[1] Thus upon incubation of Bz-Leu-OH and H-Gly-NHPh in the presence of papain, the authors failed to synthesize the desired dipeptide Bz-Leu-Gly-NHPh but obtained instead the leucine derivative Bz-Leu-NHPh. The reaction sequence was probably as follows; the enzymatic cleavage of the amide linkage of the amine component resulted in the release of free glycine and aniline, following which the latter reacted via papain-catalysis with the original carboxyl component to give Bz-Leu-NHPh. Thus, the acyl-portion (R–CO–) of one reactant, namely, the glycyl-moiety of the amine component, was replaced by a second acyl-group, i.e., the benzoylleucyl moiety. This process can be considered as a papain-induced transacylation; a phenomenon which was observed repeatedly during further experiments.[2-4] However, undesired by-products may also arise from transamidation of transpeptidation events, i.e., by an exchange of the imino-portion (–NH–R) of the carboxyl component with a different imino moiety (–NH–R'). Formally, the tryptic conversion of nuclease-T fragments to covalent forms, which was discussed in detail in Chapter 10, also corresponds to a transpeptidation process.[5] In this case the Lys_{48}-Lys_{49} peptide bond of the nuclease (6 to 49) fragment, which was originally created by tryptic hydrolysis of nuclease-T (6 to 149), is tryptically cleaved and subsequently replaced by a new peptide bond between Lys_{48} and the N-terminal Gly_{50} residue of the second hydrolysis product, i.e., the nuclease (50 to 149) fragment. The question of whether or not the excision of the Lys_{49}-residue proceeds via direct transpeptidation without releasing an intermediate free nuclease (6 to 48) fragment, was not further studied.

Truncated peptides also resulted from papain[4] (see below) and carboxypeptidase Y-mediated[6] transpeptidation reactions. Furthermore, oligomerization processes may also give rise to random insertions of additional amino acid residues into the growing peptide chain. Thus, for example, side reactions of this type were observed when Bz-Phe-OH and H-Gly-NHPh were reacted in the presence of papain, the tripeptide Bz-Phe-Gly-Gly-NHPh,[2] or, when several Boc-peptidyl-phenylhydrazides were created during the synthesis of the enkephalins. Here the molar ratio of glycine to phenylalanine residues was variously 1:2, 6:1, and 3:1 instead of the expected ratio which should read 2:1.[4] Oligomerization was also shown to occur during carboxypeptidase-Y-controlled reactions.[7] One of these processes, namely the dimerization of glycine, was actually taken advantage of for the preparation of an enkephalin-subunit,[8] however, during the CPD-Y -catalyzed semisynthesis of the B-chain of insulin, an unexpected dimerization of threonine led to a by-product containing an additional threonine unit.[9]

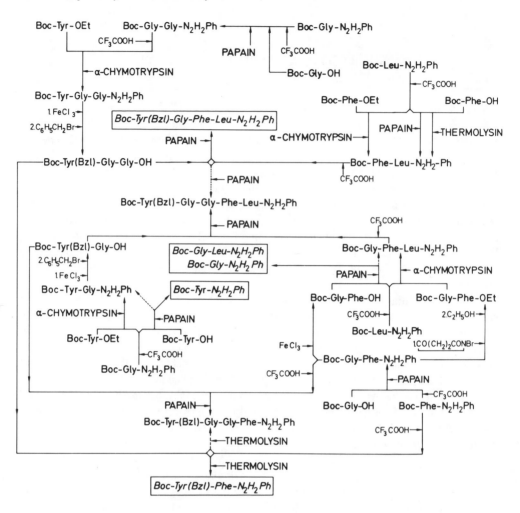

FIGURE 1.   Successful and unsuccessful pathways of the enzyme-mediated preparation of Leu-enkephalin. Dashed lines indicate projected but unobtainable products. Unwanted by-products of reactions are boxed.

   During our own studies on the protease-mediated preparation of Leu-enkephalin, several attractive synthetic concepts had to be discarded or modified, when they failed to provide the projected peptides (cf. Figure 1).[4] Here, *a priori* cleavage of pre-existing peptide bonds followed by enzymatic synthesis of new bonds connecting original educts and nascent degradation products gave rise to undesirable, often truncated peptides.

   Several synthetic pathways which had to be discarded because they proved to be dead ends, may serve as an illustration (cf. Figure 1). The initial design of a route to the protease-catalyzed synthesis of Leu-enkephalin required a final fragment condensation of Boc-Tyr(Bzl)-Gly-Gly-OH and H-Phe-Leu-N_2H_2Ph in the presence of papain. However, instead of yielding the targeted pentapeptide, this reaction resulted in the formation of the tetrapeptide Boc-Tyr(Bzl)-Gly-Phe-Leu-N_2H_2Ph. (The tetra- and pentapeptide could be distinguished by amino acid-, elemental-, and HPLC analysis).[10] Apparently, the COOH-terminal glycine residue of the tripeptide had been proteolytically removed (the release of free glycine could be detected), followed by papain-mediated peptide bond formation between the resulting di-peptide Boc-Tyr(Bzl)-Gly-OH and the original amine component, the dipeptide H-Phe-Leu-N_2H_2Ph. It was not established if the Boc-Tyr(Bzl)-Gly-papain complex was directly ami-nolyzed, i.e., whether a direct transpeptidation took place, or alternatively if the acyl-papain complex was hydrolyzed prior to the coupling of the newly emerged cleavage product Boc-Tyr(Bzl)-Gly-OH to H-Phe-Leu-N$_2$H$_2$Ph. This example demonstrates once again how either a direct or indirect (via hydrolytic cleavage) transpeptidation reaction can subvert an apparently promising synthetic design. Nevertheless, this result suggested, that the dipeptide Boc-Tyr(Bzl)-Gly-OH might be a more suitable carboxyl component for such a papain-catalyzed peptide bond formation than the tripeptide Boc-Tyr(Bzl)-Gly-Gly-OH. Indeed as already described (Chapter 8, Section II) and outlined in Figure 1, this synthetic route finally led to the desired pentapeptide.

Boc-Tyr-Gly-N$_2$H$_2$Ph, a precursor of the abovementioned dipeptide, could be synthesized via chymotrypsin-catalysis,[11] while an alternative approach to the synthesis of this Boc-dipeptidyl-phenylhydrazide via papain-controlled coupling of Boc-Tyr-OH and H-Gly-N$_2$H$_2$Ph failed because Boc-Tyr-N$_2$H$_2$Ph was formed instead (Figure 1).[4] In the light of previous comments, the "by-product" Boc-Tyr-N$_2$H$_2$Ph can be considered to result from a transacylation process, since the acyl-portion of the initial amine component, i.e., the glycyl moiety of H-Gly-N$_2$H$_2$Ph, was replaced by another acyl-group, i.e., the Boc-tyrosyl moiety. In the following example both transamidation and tranacylation steps were responsible for the unexpected consequences of papain-induced "side reactions". Thus, in order to synthesize the tripeptide Boc-Gly-Phe-Leu-N$_2$H$_2$Ph, Boc-Gly-Phe-OH and H-Leu-N$_2$H$_2$Ph were incubated together in the presence of papain. Contrary to intentions, the reaction actually yielded Boc-Gly-Leu-N$_2$H$_2$Ph and Boc-Gly-N$_2$H$_2$Ph (Figure 1). The synthetic routes leading to these unwanted compounds were probably as follows. First, one observed proteolytic cleavage of the Gly-Phe bond of the carboxyl component, and also — at least in part — of the Leu-N$_2$H$_2$Ph linkage of the amine component, following which the newly formed Boc-Gly-OH reacted both with the initial amine component to give Boc-Gly-Leu-N$_2$H$_2$Ph, and with the released phenylhydrazine to yield Boc-Gly-N$_2$H$_2$Ph. Next, since it seemed possible that the desired tripeptide sequence could be obtained through a reduction in the solubility of the prospective end product, the original educts Boc-Gly-Phe-OH and H-Leu-N$_2$H$_2$Ph were replaced by Bpoc-Gly-Phe-OH and H-Leu-OTMB, respectively. This procedure, too, proved unsuccessful. By analogy to the previously described reaction scheme, papain catalyzed the synthesis of Bpoc-Gly-Leu-OTMB. Bpoc-Gly-OTMB, however, was not formed; obviously, the Leu-OTMB bond was resistant to papain-induced cleavage.

An alternative approach to synthetic Leu-enkephalin required a final α-chymotrypsin-assisted coupling step between Boc-Tyr(Bzl)-Gly-Gly-Phe-OEt and H-Leu-N$_2$H$_2$Ph.[11] Although Boc-Tyr(Bzl)-Gly-Gly-Phe-N$_2$H$_2$Ph, a precursor of the above tetrapeptide ethyl ester, could be prepared via papain-catalysis (cf. Chapter 8, Section II), it could not be obtained through a thermolysin-controlled coupling of Boc-Tyr(Bzl)-Gly-Gly-OH and H-Phe-N$_2$H$_2$Ph (Figure 1). In fact, the thermolysin-catalyzed reaction produced the dipeptide Boc-Tyr(Bzl)-Phe-N$_2$H$_2$Ph. According to analytical data, the Tyr-Gly bond of the tripeptide was hydrolyzed at the beginning of the reaction to release Boc-Tyr(Bzl)-OH and the free dipeptide H-Gly-Gly-OH. The first-mentioned cleavage product then reacted with the initial amine component H-Phe-N$_2$H$_2$Ph to yield the dipeptide Boc-Tyr((Bzl)-Phe-N$_2$H$_2$Ph.

The above descriptions serve to illustrate the unpredictable nature of many of the protease-catalyzed reactions attempted during the preparation of Leu-enkephalin. Clearly, a critical evaluation of the chemical nature of the resultant peptides is indispensable to avoid serious confusions. A rapid preliminary characterization of the reaction products by a combination of thin-layer chromatography and various staining methods employing group-specific chromogenic reagents allows a convenient and reliable judgement on the success or failure of a synthetic pathway.[12] In cases where these procedures did not provide unambiguous conclusions, HPLC techniques have proved to be valuable alternatives.[10] A posteriori, those provisional analyses could always be confirmed by amino acid and elemental analyses.

In summary, the uncertainties in the outcome of protease-catalyzed reactions presently limit the general applicability of the enzymatic approach to peptide synthetic chemistry. Currently, there does not exist a generally valid method to suppress undesired side reactions mediated by proteases. However, in most instances the reaction pathways which prove to be dead ends can be bypassed by alternative strategies. Under these circumstances, one should adopt a versatile, synthetic design which then allows for subsequent improvisations. It is self-evident, that the ideal concept would enable the synthesis of the target peptide by a concerted action of proteases with nonoverlapping specificities. In cases where this perfect state of affairs cannot be achieved, for example where the preparation of the desired peptide requires the repeated use of one protease, individual ''subunits'' can first be synthesized separately in the presence of one protease. The fragments thus obtained are then assembled in the presence of another protease that does not endanger the preformed peptide bonds to yield the final completed peptide. Even the same protease can sometimes be used more than once during the stepwise synthesis of a peptide; but only where the rate of peptide bond formation markedly exceeds the rate of proteolytic cleavage of the pre-existing, ''scissile'' bond. By resorting to one or the other of these expedients many of the problems arising during various enzymatic syntheses can be overcome.[13-15] In principle, an exopeptidase such as carboxypeptidase Y ought be the optimal choice to catalyze a stepwise synthesis without endangering internal peptide bonds. However, as was observed during the preparation of semisynthetic insulin,[9] detrimental side reactions such as transamidation- or oligomerization processes may also occur in the presence of carboxypeptidase Y.

A novel and interesting variant of enzymatic peptide synthetic methodology was suggested by Nakajima et al.[16] These authors used aminoacyl-*t*-RNA synthetases, enzymes which are involved in ribosome-mediated protein synthesis and which demonstrate high specificity for their cognate amino acids, to prepare a series of dipeptides. The specificity of the synthetases was of course confined to the carboxyl components which were used in a pre-activated state, i.e., in form of the respective aminoacyl adenylates. Whether such an approach can be applied on a preparative scale remains to be seen. Because the synthetases exhibit specificity solely for aminoacyl- and not for peptidyl adenylates, peptides would be required to play the part of amine components if the peptide chains were to be extended beyond the level of dipeptides. This would become a rather expensive procedure given the cost of the synthetases and the large excess of amine component necessarily used during the above study.

If nature cannot supply the peptide synthetic chemist with an adequate set of proteases one might proceed according to the motto: ''Need a catalyst? Design an enzyme''.[17] A variety of proteinaceous and nonproteinaceous compounds have been prepared in an effort to mimic the catalytic action of proteases. (The following references only give a limited survey of this challenging field of enzyme chemistry, 18 to 27.) However, the studies to date on protease mimetics have focused predominantly on their hydrolytic activities, while few reports on enzyme models displaying proteosynthetic capacities have been published. In this respect, the work of Sasaki et al.[28] on the preparation of an artificial catalyst for the synthesis of peptide bonds represents a rare exception. These authors used a crown ether as scaffold to which two thiol-groups were fixed as catalytic functions. The artificial enzyme thus resembled a miniature organic model of the antibiotic synthetases which function as catalysts in nonribosomal peptide biosynthesis.[29] The educts, i.e., the carboxyl- and the amine component, were covalently linked by chemical means to the enzyme mimic via thioester bonds (Figure 2). Intramolecular aminolysis resulted in the formation of a peptide bond with the growing peptide chain still bound to the carrier through a thioester linkage. After successive rounds of amino acid addition the completed tetrapeptide was finally cleaved from the crown ether by methanolysis. Although the binding of the substrates to the enzyme mimetic involved purely chemical steps, the respective peptide-bond-forming processes can be regarded as being enzyme-catalyzed. Consequently, the outcome of this study is encouraging and should stimulate further efforts in this field.

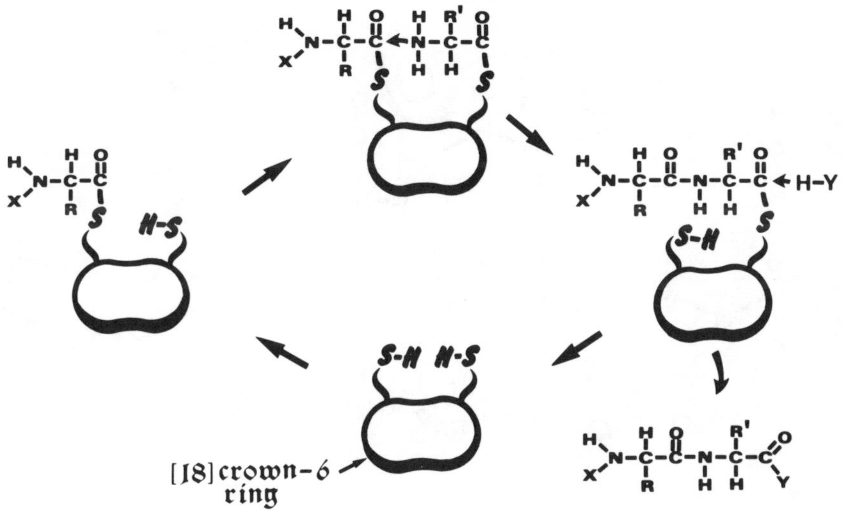

FIGURE 2.   Peptide bond formation mediated by an enzyme-mimicking functionalized crown ether.

As mentioned previously, the so-called peptidyltransferase, a ribosomal protein that ca-talyzes the peptide-bond formation in vivo should be an ideal choice to serve this function in vitro as well. This enzyme is able to catalyze the incorporation of all the codogenous amino acid residues into a growing peptide chain in a stereo- and regiospecific manner. Thus in principle, the peptidyltransferase meets all the requirements for enzyme-mediated peptide synthesis. However, its proteosynthetic activity is dependent upon the presence of other ribosomal "helper" proteins[30] and consequently one can hardly expect that, when isolated from its natural environment, the enzyme can fully preserve its original functions.

Since it is unlikely that an autonomous, fully operative *ex natura* peptidyltransferase will be available within a reasonable space of time, attempts have been made to design and synthesize *ex arte* peptidyltransferases.[31] These artificial enzymes have been designed both to accept as substrates the largest possible number of amino acid residues and to display their catalytic activities in organic solvents in order to favor proteosynthesis at the expense of proteolysis. The activity design of the presumptive synthetase mimetic peptide, which determined the relevant catalytic groups and their disposition in space,[32] was based upon the active site of serine proteases represented by the triad Asp-Ser-His (cf. Chapter 4, Figure 3).[33] These amino acid residues, the side-chain functions of which constitute a proton transfer system along hydrogen bonds, were integral parts of cyclosymmetric decapeptides (Figure 3). Conformationally cyclic peptides are less flexible than their linear counterparts and their relatively rigid shape not only improves the prospects of suitably positioned binding in relation to catalytic groups but also prevents the complete randomization of the peptidic backbone in the presence of denaturing agents. Consequently, cyclopeptides should be more resistant to the destabilizing effects of organic solvents which are known to reduce the biological activities of naturally occurring enzymes.

The artificial "peptide synthetases" consisted of two identical pentapeptides the head-to-tail cyclization of which was favored by the incorporation of both L- and D-amino acids. The decapeptide cylco-(-Ser-D-His-Asp-D-Phe-Pro-)$_2$ and its mirror image were assembled in a bidirectional manner by the solid-phasé method. The synthesis was based upon an orthogonal protecting scheme[34] which involved the use of acid- and base-labile as well as thiolate-sensitive blocking groups. According to preliminary results, the cyclic decapeptides are capable of stereospecifically catalyzing the synthesis of model peptides. For example the synthetase mimic, cyclo-(-D-Ser-His-D-Asp-Phe-D-Pro-)$_2$, (Figure 3b) demonstrated a

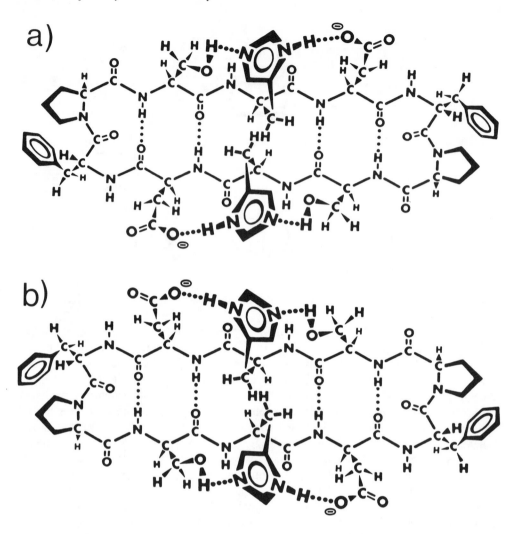

FIGURE 3.   Proposed molecular structure of artificial synthetases: (a) cyclo-(-Ser-D-His-Asp-D-Phe-Pro-)$_2$; (b) cyclo-(-D-Ser-His-D-Asp-Phe-D-Pro-)$_2$.

strong preference for L-amino acids in the P$_1$-position of its substrates. Conversely, the enantiomeric form, cyclo-(-Ser-D-His-Asp-D-Phe-Pro-)$_2$, (Figure 3a) favored D-amino acids in the P$_2$-site over L-amino acids. Thus, the synthesis of the enkephalin precursor, Boc-Tyr-Gly-N$_2$H$_2$Ph (cf. Figure 1), was preferentially catalyzed by the decapeptide enantiomer shown in Figure 3b, whereas the synthesis of the mirror image, Boc-D-Tyr-Gly-N$_2$H$_2$Ph, was better catalyzed by the enantiomer shown in Figure 3a. These results may pave the way for the enzymatic incorporation not only of L-amino acids but possibly also of their optical antipodes, i.e., of the D-amino acids, into target peptides; an objective that could hardly be attained by the use of natural enzymes with their strict specificity for L-amino acids.

The suggestion:[17] "Need a catalyst? Design an enzyme", can also be realized by using recombinant DNA techniques instead of chemical *de novo* syntheses of the type outlined above. Just to give one example relevant to proteases; Clark et al. succeeded in redesigning trypsin via site-directed mutagenesis to obtain a modified trypsin molecule with altered substrate specificity.[35]

An additional drawback of the enzymatic approach to peptide synthesis results from the necessity to use relatively high concentrations not only of educts but also of enzymes to obtain reasonably rapid rates of peptide bond formation. According to Fruton,[36] this short-coming can be attributed to the unfavorable equilibrium of the condensation reaction

$$RCOO^- + {}^+NH_3R' + EH \rightleftharpoons [RCO]E[NH_2R'] + H_2O$$

These disadvantages may be ameliorated to some extent by using immobilized proteases which can be successfully reutilized. An alternative is the so-called iterative procedure proposed by Petkov and Stoineva.[37] In their study on the synthesis of protected dipeptides the authors used "mobile" proteases (chymotrypsin and thermolysin) and a large excess of the amine component. (In particular the preparation of Z-Tyr-D-Leu-NH$_2$ by chymotryptic catalysis from Z-Tyr-OMe and H-D-Leu-NH$_2$ requires a high concentration of the amine component, because H-D-Leu-NH$_2$ is a rather poor nucleophile). After completion of the first reaction the precipitated products were removed and a further round of synthesis was started by adding equimolar amounts of carboxyl- and amine components. This procedure, which could be performed repeatedly, enabled efficient usage not only of the proteases but also of the surplus fraction of the amine component. Furthermore, reaction media which often contain water-miscible organic co-solvents may cause denaturation of the proteases. The consequences of this effect can be obviated by increasing the concentration of the enzyme. However, a more elegant alternative is the use of biphasic systems, which largely prevent the enzymes from coming into unfavorable contacts with organic co-solvents (cf. Chapter 5, Section II.A).

REFERENCES

1. **Bergmann, M. and Fraenkel-Conrat, H.,** The enzymatic synthesis of peptide bonds, *J. Biol. Chem.,* 124, 1, 1938.
2. **Janssen, F., Winitz, M., and Fox, S. W.,** Enzymic synthesis of peptide bonds. V. Instances of protease-controlled specificity in the synthesis of acylamino acid anilides and acylpeptide anilides, *J. Am. Chem. Soc.,* 75, 704, 1953.
3. **Milne, B. H. and Most, Jr., C. F.,** Peptide synthesis via oxidation of N-acyl-α-amino acid phenylhydrazides. II. Benzyloxycarbonyl peptide phenylhydrazides, *J. Org. Chem.,* 33, 169, 1968.
4. **Kullmann, W.,** Protease-mediated peptide bond formation. On some unexpected outcomes during enzymatic synthesis of Leu-enkephalin, *J. Biol. Chem.,* 256, 1301, 1981.
5. **Homandberg, G. A. and Chaiken, I. M.,** Trypsin-catalyzed conversion of Staphylococcal nuclease-T fragment complexes to covalent forms, *J. Biol. Chem.,* 255, 4903, 1980.
6. **Breddam, K., Widmer, F., and Johansen, J. T.,** Carboxypeptidase Y catalyzed transpeptidations and enzymatic peptide synthesis, *Carlsberg Res. Commun.,* 45, 237, 1980.
7. **Widmer, F., Breddam, K., and Johansen, J. T.,** Carboxypeptidase Y catalyzed peptide synthesis using amino acid alkyl esters as amine components, *Carlsberg Res. Commun.,* 45, 453, 1980.
8. **Widmer, F., Breddam, K., and Johansen, J. T.,** Carboxypeptidase Y as a catalyst for peptide synthesis in aqueous phase with minimal protection, in *Peptides 1980, Proc. 16th Eur. Peptide Symp.,* Brunfeldt, K., Ed., Scriptor, Copenhagen, 1981, 46.
9. **Breddam, K., Widmer, F., and Johansen, J. T.,** Carboxypeptidase Y catalyzed C-terminal modification in the B-chain of porcine insulin, *Carlsberg Res. Commun.,* 46, 361, 1981.
10. **Kullmann, W.,** Monitoring of protease-catalyzed peptide synthesis by high performance liquid chromatography, *J. Liquid Chromatog.,* 4, 1121, 1981.
11. **Kullmann, W.,** Proteases as catalysts for enzymic syntheses of opioid peptides, *J. Biol. Chem.,* 255, 8234, 1980.
12. **Kullmann, W.,** Rapid characterization by thin-layer chromatography of amino acid and peptide derivatives enzymically prepared during protease-mediated peptide synthesis, *J. Liquid Chromatog.,* 4, 1947, 1981.
13. **Kullmann, W.,** Enzymatic synthesis of dynorphin$_{1-8}$, *J. Org. Chem.,* 47, 5300, 1982.
14. **Kullmann, W.,** Protease-catalyzed peptide bond formation: Application to synthesis of the COOH-terminal octapeptide of cholecystokinin, *Proc. Natl. Acad. Sci. U.S.A.,* 79, 2840, 1982.

15. **Kullmann, W.,** Protease-catalyzed synthesis of melanocyte-stimulating hormone (MSH) fragments, *J. Protein Chem.,* 2, 289, 1983.

16. **Nakajima, H., Kitabatake, S., Tsurutani, R., Tomioka, I., Yamamoto, K., and Imahori, K.,** Reactions of the aminoacyl-tRNA synthetase-aminoacyl adenylate complex and amino acid derivatives. A new approach to peptide synthesis, *Biochim. Biophys. Acta,* 790, 197, 1984.

17. **Maugh, II, T. H.,** Need a catalyst? Design an enzyme, *Science,* 223, 269, 1984.

18. **Sheehan, J. C., Bennett, G. B., and Schneider, J. A.,** Synthetic peptide models of enzyme active sites. III. Stereoselective esterase models, *J. Am. Chem. Soc.,* 88, 3455, 1966.

19. **Petz, D., and Schneider, F.,** Synthesis and catalytic properties of peptides with hydrolytic activity, *Z. Naturforsch.,* 31c, 534, 1976.

20. **Nakajima, B.-I. and Nishi, N.,** Synthesis of linear-, cyclic-, and poly-peptides having the sequence -Asp-εAhx-Ser-εAhx-His-εAhx-, and their ester hydrolytic reactions, in *Peptide Chemistry 1982,* S. Sakakibara, Ed., Protein Research Foundation, Osaka, Japan, 1983, 41.

21. **Cram, D. J. and Katz, H. E.,** An incremental approach to hosts that mimic serine proteases, *J. Am. Chem. Soc.,* 105, 135, 1983.

22. **Breslow, R., Trainor, G., and Ueno, A.,** Optimization of metallocene substrates for β-cyclodextrin reactions, *J. Am. Chem. Soc.,* 105, 2739, 1983.

23. **Mallick, J. M., D'Souza, V. T., Yamaguchi, M., Lee, J., Chalabi, P., Gadwood, R. C., and Bender, M. L.,** An organic chemical model of the acyl-α-chymotrypsin intermediate, *J. Am. Chem. Soc.,* 106, 7252, 1984.

24. **Masuda, Y., Tanihara, M., Imanishi, Y., and Higashimura, T.,** Hydrolysis of active esters of aliphatic carboxylic acids with cyclic dipeptide catalysts consisting of L-histidine and different aliphatic α-amino acids, *Bull. Chem. Soc. Jpn.,* 58, 497, 1985.

25. **Skorey, K. J. and Brown, R. S.,** Biomimetic models for cysteine proteases. II. Nucleophilic thiolate-containing zwitterions produced from imidazole-thiol pairs. A model of the acylation step in papain-mediated hydrolyses, *J. Am. Chem. Soc.,* 107, 4070, 1985.

26. **D'Souza, V. T., Hanabusa, K., O'Leary, T., Gadwood, R. C., and Bender, M. L.,** Synthesis and evaluation of a miniature organic model of chymotrypsin, *Biochem. Biophys. Res. Commun.,* 129, 727, 1985.

27. **Menger, F. M. and Whitesell, L. G.,** A protease mimic with turnover capabilities, *J. Am. Chem. Soc.,* 107, 707, 1985.

28. **Sasaki, S., Shionoya, M., and Koga, K.,** Functionalized crown ethers as an approach to the enzyme model for the synthesis of peptides, *J. Am. Chem. Soc.,* 107, 3371, 1985.

29. **Lipmann, F.,** Nonribosomal polypeptide synthesis on polyenzyme templates, *Acc. Chem. Res.,* 6, 361, 1973.

30. **Hampl, H., Schulze, H., and Nierhaus, K. H.,** Ribosomal components from *Escherichia coli* 50 S subunits involved in reconstitution of peptidyltransferase activity, *J. Biol. Chem.,* 256, 2284, 1981.

31. **Kullmann, W.,** Design and synthesis of peptide synthetase mimics, in preparation.

32. **Robson, B.,** The design of biologically active polypeptides, *Crit. Rev. Biochem.,* 14, 273, 1984.

33. **Blow, D. M.,** Structure and mechanism of chymotrypsin, *Acc. Chem. Res.* 9, 145, 1976.

34. **Barany, G. and Merrifield, R. B.,** A new amino protecting group removable by reduction. Chemistry of the dithiasuccinoyl (Dts) function, *J. Am. Chem. Soc.,* 99, 7363, 1977.

35. **Craik, C. S., Largman, C., Fletcher, T., Roczniak, S., Barr, P. J., Fletterick, R., and Rutter, W. J.,** Redesigning trypsin: alteration of substrate specificity, *Science,* 228, 291, 1985.

36. **Fruton, J. S.,** Proteinase-catalyzed synthesis of peptide bonds, *Adv. Enzymol. Relat. Areas Mol. Biol.,* 53, 239, 1982.

37. **Petkov, D. D. and Stoineva, I. B.,** Enzyme peptide synthesis by an iterative procedure in a nucleophile pool, *Tetrahedron Lett.,* 25, 3751, 1984

Chapter 14

SYNOPSIS AND PROGNOSIS

The extraordinary capabilities inherent in biocatalysts have long evoked the interest of organic chemists. In comparison to common chemical catalysts, enzymes are able to catalyze reactions with unparalleled efficiency and with unique substrate specificity under the mildest conceivable conditions. Considering these favorable features, it comes as no surprise that enzymes are widely exploited in various branches of science and engineering. Hence, the peptide synthetic chemist is also eagerly bent on exploiting the properties of those enzymes which best satisfy his peculiar purposes, namely, the proteases.

A superficial view of the thermodynamic aspects of protease- catalyzed processes might suggest that the enzymatic approach to peptide synthetic chemistry is not feasible. Nevertheless, many studies on enzymatic syntheses of both model and naturally occurring peptides, as well as peptide and protein semisyntheses have demonstrated the capabilties of various proteases as proteosynthetic catalysts.[1-5] However, despite these successes, the protease-based deficiencies of enzymatic peptide synthesis cannot be ignored. Although the enzymatic approach to peptide synthetic chemistry may be described as being rapidly developing, it is still in its infancy and has not yet reached the versatility of a standard procedure. However, these shortcomings are likely to be overcome by a systematic evaluation of the proteosynthetic potentials of already known enzymes and also by a methodical search for new proteases, especially those from microbial origin. Impetus for further research may be provided by the statement of Laskowski, Jr.:[6] "The continual stream of discoveries of new proteinases with very strict specificity gives me a great deal of hope." Beyond that, artificial catalysts which are able to mimic the proteosynthetic capacities of proteases may be constructed, either by creating totally new enzymes or by redesigning natural enzymes.

Given the present "state-of-the-art", protease-mediated peptide synthesis should be considered as a supplement, rather than an alternative, to chemical procedures. At present, the enzymatic method may be best exploited for the total synthesis of smaller peptides, such as neuropeptides or their "mutated" analogs, for synthetic strategies combining enzymatic as well as chemical techniques and for semisyntheses of larger peptides and proteins. It will require further work to establish sophisticated, multi-step enzyme systems the components of which would be able to operate in a sequential order to catalyze the individual chemical steps involved in peptide synthesis. This would include in addition to actual peptide bond formation, the incorporation and removal of protecting groups. The integration of these procedures would represent a second generation enzyme technology. Whether, in the future, enzyme-catalyzed peptide synthesis will be integrated as a fully autonomous technique into peptide synthetic methodology remains to be seen.

REFERENCES

1. **Chaiken, I. M., Komoriya, A., Ohno, M., and Widmer, F.,** Use of enzymes in peptide synthesis, *Appl. Biochem. Biotechnol.,* 7, 385, 1982.
2. **Fruton, J. S.,** Proteinase-catalyzed synthesis of peptide bonds, *Adv. Enzymol. Relat. Areas Mol. Biol.,* 53, 239, 1982.
3. **Jakubke, H.-D. and Kuhl, P.,** Proteasen als Biokatalysatoren für die Peptidsynthese, *Pharmazie,* 37, 89, 1982.
4. **Jakubke, H.-D., Kuhl, P. and Könnecke, A.,** Grundprinzipien der proteasekatalysierten Knüpfung der Peptidbindung, *Angew. Chem.,* 97, 79, 1985.

5. **Kullmann, W.,** Proteases as catalysts in peptide synthetic chemistry. Shifting the extent of peptide bond synthesis from a "quantité négligeable" to a "quantité considérable", *J. Prot. Chem.,* 4, 1, 1985.

6. **Laskowski, M., Jr.,** The use of proteolytic enzymes for the synthesis of specific peptide bonds in globular proteins, in *Semisynthetic Peptides and Protein,* Offord, R. E. and DiBello, C., Eds., Academic Press, New York, 1978, 255.

ABBREVIATIONS

Amino acids are of the L-configuration unless otherwise indicated. Symbols used in this volume are listed below (except those of the common amino acids).

AA =	amino acids
Ac =	acetyl
Acm =	acetamidomethyl
Bz =	benzoyl
Bzl =	benzyl
Z =	benzyloxycarbonyl
Bpoc =	2-(p-biphenyl)propyl(2)oxycarbonyl
Boc =	tert-butyloxycarbonyl
Bu^t =	tert-butyl
OBu^t =	tert-butyl ester
OBt =	benzotriazolyl ester
CD =	circular dichroic spectroscopy
ODPM =	diphenylmethylester
DCC =	dicyclohexylcarbodiimide
Dnp =	2,4-dinitrophenyl
OEt =	ethyl ester
Fmoc =	9-fluorenylmethyloxycarbonyl
⌐Glu =	pyroglutamic acid
HPLC =	high-performance liquid chromatography
HOBt =	*N*-hydroxybenzotriazole
HOSu =	*N*-hydroxysuccinimide
Z(OMe) =	4-methoxybenzyloxycarbonyl
OMe =	methyl ester
Msc =	methylsulfonylethyloxycarbonyl
pNA =	*p*-nitroanilide
Ph =	phenyl
OSu =	succinimidyl ester
tlc =	thin-layer chromatography
TFA =	trifuoracetic acid
OTMB =	2,4,6-trimethylbenzyl ester

INDEX

T

Tacynase I, 56
Temperature, effects on equilibrium, 24, 30—31
Template, polynucleotide, 6
Tertiary structure, of chymotrypsin, 19
Thermitase, 49, 121
Thermoactinomyces vulgaris, 49, 121
Thermodynamic aspects of peptide bond synthesis,
    9—12
Thermolysin, 129
    enzymatic synthesis, 67—77
    formation of peptide bonds, 20
    kinetics, 106
    proteases as biocatalysts for synthesis of model
        peptides, 45, 53—56
    protecting group chemistry, 122
Thio-ester, activated, 20
Threonine dimerization, 127
Thrombin, 100
Transacylation, 127, 129
Transamidation, 41, 127, 129
Transition state, 13—18
Transmission coefficient, 14

Transpeptidation, 107
    biologically active peptides, 65
    carboxypeptidase Y, 49—50
    protease-catalyzed oligomerization, 83
    protease-catalyzed semisynthesis, 89—90
    shortcomings and alternatives, 127—129
Tritiachium album, 95
Trypsin
    catalytic pathway of, 20
    ester cleavage by, 120—121
    in plastein formation, 5
    semisyntheses with, 88—92, 98—100, 107
    syntheses with, 45—47, 69, 71, 74—76
Trypsin-kallikrein inhibitor, 94—95, 106, 124

V

Vasopressin, deamino-lysine, 124

Z

Zwitterionic reactants, 9, 23, 107
Zymo-hydrolysis, reversible, 5—6